Managing a Changing Climate in Africa

Local Level Vulnerabilities and Adaptation Experiences

Pius Z. Yanda
Chipo P. Mubaya

Mkuki na Nyota
DAR – ES – SALAAM

Published by:

Mkuki na Nyota Publishers Ltd.

Nyerere Road, Quality Plaza Building

P. O. Box 4246

Dar es Salaam, Tanzania

www.mkukinanyota.com

For:

Institute of Resource Assessment (IRA)

University of Dar Es Salaam

P. o. Box 35097

Dar Es Salaam

ww.ira.udsm.ac.tz

Photo Credit:

Maua Seminary community

Christine Cyprian

Mkuki Bgoya

Deo Simba

Authors

ISBN 978-9987-08-089-2

Contents

PART 1

PART 2

PART 3
Challenges and Opportunities in Addressing Climate Change Impacts

List of Tables

List of Figures

Acknowledgement

Assistance with literature searches for this book came from Mr. Edmund Mabhuya. The authors acknowledge funding for case studies from the Climate Change Adaptation in Africa (CCAA) programme, START and CODESRIA. The authors would also like to acknowledge collaborators and researchers from Zambia Agricultural Research Institute (ZARI), Zambia Agro-Met, Midlands State University (MSU) Zimbabwe, ICRISAT Zimbabwe and CIAT Zimbabwe for the case studies from Southern Africa.

Authors

Pius Zebhe Yanda is a Professor of Physical Geography at the Institute of Resource Assessment, University of Dar Es Salaam. He has served the University of Dar Es Salaam in different academic positions for 24 years. Since 2006 he has been serving as the Director of the Institute of Resource Assessment. Professor Yanda has recently been appointed to chair the newly established Mwalimu Julius Nyerere Professorial Chair in Environment and Climate Change at the University of Dar Es Salaam. He has been involved in various programmes, the most recent being: (i) regional training programme on climate change and biodiversity conservation, with participants from DRC, Burundi, Rwanda, Uganda and Tanzania; and (ii) fellowship programme involving four types of fellowships (post-doctoral, doctoral, teaching and policy) for the period ranging from six months to one year.

Dr. Chipo Plaxedes Mubaya's research background is Natural Resource Management. Specific focus has been on policy and institutional arrangements, livelihood assessments and issues to do with access and use of these resources. Chipo also has experience in understanding gender dimensions in Community-Based Natural Resource Management such as water and forests, among others. She also has experience in Integrated Agricultural Research for Development and enabling rural innovations through linking farmers to markets, understanding social dynamics of farmers and climate change adaptation in participatory action research in Southern Africa. She is currently working with the Pan-African START Secretariat (PASS) at the University of Dar es Salaam capacity building and research on climate change adaptation and mitigation.

Introduction

The realisation that the Earth's climate might be sensitive to the atmospheric concentrations of gases that create a greenhouse effect is more than a century old (IISD 2008; IPCC[1] 2007d: 7-9). Scientists such as Fourier (French) and Arrhenius (Swedish) explained the Earth's greenhouse effect and the role played by some atmospheric gases such as CO_2 and methane (CH_4) in warming our planet (Fleming, 1998). Around the same time, Arrhenius, together with Chamberlain, an American scientist, realised that the burning of fossil fuels could lead to global warming. Indeed, there is increasing evidence from work that has been carried out over nearly two decades by the IPCC, which cements the conclusion that global warming and the subsequent climate changes are largely due to human activities. However, there continues to be considerable debate regarding the causes of climate change, that is, whether it is induced by anthropogenic activities or simply within the range of natural variability in climate. While some scholars theorise that climate change is largely due to natural changes, others postulate that anthropogenic factors are the major cause. Suffice it to say at this point in the debate there appears to be consensus regarding the fact that climate change is set to accelerate, calling for the need to employ strategies to deal with these changes.

The historical climate record for Africa and the rest of the world indicate higher temperatures during the 20th century, changes in precipitation, a rise in the sea level and in the extent of ice and snow. Climate change associated with increasing levels of carbon dioxide is likely to affect developed and developing countries differently, with major vulnerabilities occurring in low-latitude regions. For Africa, there has been a warming of approximately 0.7°C across most of the continent in the 20th century. Although these warming trends seem to be the same over the African continent, climate changes are not always uniform. This provides impetus for research activities that endeavour to capture trends in climate change across Africa in order to inform policies that are geared towards addressing problems emanating from these changes.

---

1 The IPCC was established by the World Meteorological Office (WMO) and UNEP in 1988. Its mandates included identification of gaps in knowledge on climate changes and potential impacts, identification of relevant information for evaluation of policy implications. The IPCC was also tasked to review planned international policies that deal with GHG issues and to assess these policies and make recommendations to governments and NGOs for socio-economic and environmental development (IPCC 2007).

Climate variability directly affects agricultural production since agriculture is inherently sensitive to climate conditions and is one of the most vulnerable sectors to the risks and impacts of global climate change. Therefore, concern for future impacts on agriculture caused by changes in important natural resources, especially land and water, seems to be justified. Climate changes also affect crops and livestock as pests and disease infestation is exacerbated in warmer and humid climates. Additional use of chemicals for both soil and livestock may further impact water and air quality. Be that as it may, it is important to note that climate change amplifies already existing risks for farmers. This is the case as farmers are also confronted with non-climatic risk factors such as economic instability, trade liberalisation and poor governance, among others. It is against this background that this book is divided into three major parts.

Part One, 'Climate Change in Africa: An Overview' highlights the conceptual framework of the book. This framework is itself divided into four stand alone chapters; Chapter One describes the context of climate change in Africa. More specifically, it presents climate change and how it has impacted rural communities in Africa; it introduces key concepts in the context of climate change and introduces the politics surrounding greenhouse gas emissions from developed and developing countries. The conclusion highlights the parties responsible for mitigation and adaptation to climate change. Chapter Two highlights the impact of, and vulnerability and adaptation to climate change, with specific focus on Africa. In this chapter, the authors distinguish between coping and adaptation, present a synergistic approach to these response mechanisms and outline sustainable development issues in the context of adaptation to climate change. Furthermore, this chapter briefly presents challenges to adaptation across spatial and temporal scales in Africa. Chapter Three gives an exposition of the financial requirements for countries to reduce greenhouse gases, for technological transfer (acceleration of technological advancement that encourages low emission rates) and funding mechanisms for technology transfer in developing countries. The authors conclude this chapter by presenting challenges associated with compliance to these financial and other stipulations to adaptation and by proposing a way forward regarding the issue under discussion. The last chapter of this first part, Chapter Four provides an overview of climate change and natural resources. This chapter highlights impacts of climate change on natural resources, and the role played by forests in climate change mitigation. The implications of adaptation on natural resources are outlined.

In Part Two of this book, the authors present Case Studies of Vulnerability to Climate Change and Variability in Eastern and Southern Africa. Chapter Five outlines farming community experiences of impacts from and adaptation to extreme events such as droughts and floods in Kenya, Uganda and Tanzania. These experiences are centred on impacts from both floods and droughts and how rural communities responded to these impacts. Similarly, Chapter Six highlights concrete experiences of farmers by presenting in-depth case studies from Zimbabwe and Zambia. These case studies present farmers' perceptions of climate change and variability, their experiences of impacts and how they responded to them. In contrast, Chapter Seven outlines crop production adaptation to climate change in East Africa. The chapter focuses on impacts on crop production in this region, narrowing its focus to Uganda, Kenya and Tanzania. The authors conclude this chapter by discussing major challenges to crop production.

The final section, Part Three, presents Challenges and Opportunities to Climate Change Adaptation in Africa in three chapters. Chapter Eight outlines factors that influence choice of response strategies to climate change in Zimbabwe and Zambia. This chapter presents responses strategies that farmers employed and then analyses the factors that influenced the choice of these strategies, hence highlighting challenges and opportunities that are found in these responses. Chapter Nine presents challenges to and opportunities for ecosystem-based approaches to climate change in Africa. The last chapter gives a brief presentation of challenges and opportunities arising from use of bio fuels as a mitigation measure to climate change.

PART

1

CLIMATE CHANGE IN AFRICA: AN OVERVIEW

CLIMATE CHANGE IN THE AFRICAN CONTEXT

1.0 Introduction

There is merit in the proposition that the root of climate change is global warming caused by anthropogenic emissions of carbon dioxide (CO_2), methane and other greenhouse gases (Collier *et al.* 2008). IPCC (2007) concluded that it is more than 90% certain that the current global warming is the result of human activities, particularly related to industrial consumption and land-use practices. The world has warmed by an average of 0.76°C since pre-industrial times and the global average temperature is projected to further increase by 1.8°C to 4°C if no action is taken. For example, in the Arctic average temperatures have increased at almost twice the global average rate in the past 100 years; sea ice coverage has shrunk and temperatures at the top of the permafrost layer have generally increased (IISD 2008; IPCC 2007d: 7–9).

Ultimately, the increasing temperatures, greenhouse gas accumulation in the global atmosphere and increasing regional concentrations of aerosol particles are now understood to have detectable effects on the global climate system (Santer *et al.* 1996). While these effects are evident at regional scales, they are differentiated by the technological and economic power of respective regions (Giorgi & Francisco 2000; Hulme 2001; Mitchell & Hulme 1999). The changing climate at the global scale has been considered to be more sensitive due to numerous human activities, which are directly linked to climate.

Generally, over the past century, the Earth's average surface temperature has increased by almost 0.74°C (IPCC 2007a). The consequences of this alteration are starting to become more visible as climatic conditions and ecosystems begin to change. This warming trend is projected to continue, rising another 1.1°C to 6.4°C over the next 100 years (IPCC 2007a). At present emission rates, a 2°C rise in temperature is highly probable and possibly inevitable (Stern 2006). At this level of global average temperature increase, up to 30% of all plant and animal species will likely be at an increasing risk of extinction; most corals will likely be bleached; cereal productivity in low latitudes is likely to decrease and millions of people will likely experience coastal flooding (IPCC 2007b; IISD 2008).

The climate change landscape has fundamentally altered since the Kyoto Protocol[2] which was drafted in 1997. Human-induced climate change is

2 The Kyoto Protocol (1997) is an agreement to a 5.2 % reduction in greenhouse gas emissions by about 2010 (relative to 1990), and constant emissions thereafter. These targets relate to the annex I countries. These are 38 highly industrialized countries and countries undergoing the process of transition to a market economy.

increasingly being observed (IPCC 2007a) and there is greater confidence in long-term climate projections suggesting that significant, and largely adverse change will take place within this century (IPCC 2007b). In addition, the growing number of extreme weather events throughout the world in recent years has increased sensitivity to the potentially dramatic social and economic impacts of climate change for all countries. In this respect, economic analyses have raised awareness of the substantive additional technological and financial need to prepare for these impacts (IISD 2008). In addition, these climate change trends call for ambitious local and global mitigation and adaptation efforts to reduce impacts and improve the adaptive capacity of local communities, who are most vulnerable to these impacts.

1.1 Greenhouse gas (GHG) emissions

Africa has hitherto made little contribution to the stock of greenhouse gases in the atmosphere. Data for per capita emissions of carbon dioxide, excluding land-use change, indicate that in most African countries emissions are less than 0.5 tonnes per capita. This is equivalent to one-twentieth that of the United Kingdom (Collier *et al.* 2008). Surprisingly, Sub-Saharan Africa, with 11% of the world's population, accounts for just 3.6% of world emissions of carbon dioxide, reflecting low levels of income and of energy consumption (ibid). Although current estimates indicate variations between countries, there is evidence to show that most developing countries will become significant polluters in the near future. Some of the developing countries already contributing to greenhouse gas emissions include Brazil, India, Indonesia, China, South Korea, South Africa and Mexico. This has been due to scientific and economic evolutions taking place as they attempt to compete economically with developed countries. These countries have experienced phenomenal economic growth, which has been matched by a rise in aggregate GHG emissions (IISD 2008).

For instance, China is reported to have surpassed the United States in total emissions in 2006 (Netherlands Environmental Assessment Agency 2007; IISD 2008). In addition, China alone is close to surpassing the USA in terms of emission rates and is projected to be responsible for almost 40% of global increases in emissions between 2004 and 2030 (IEA 2007: 81; IIED 2008). Overall, Brazil, India, Indonesia, China, South Korea, South Africa and Mexico in 2005 had carbon dioxide emissions equivalent to over 90% of the top five Annex I emitters (see Table 1) (IISD 2008). By 2012, if current trends continue, developing countries as a whole will overtake the OECD[3] (mainly

3 OECD: Organisation for Economic Co-operation and Development.

developed countries) as global emitters of carbon dioxide, with China and India contributing the lion's share.

However, compared with many Annex I parties, the major developing country emitters still have significantly lower economic indicators and commensurately lower GHG emissions per capita (see Table 1). Moreover, much of the rest of the developing world is still in the same position in relation to the OECD countries as they were when Kyoto was negotiated. Notwithstanding, the world will be much different again in 2012, after three more years of economic growth. In this respect, an effective post-2012 period will require flexibility to be able to account for such changes (IISD 2008).

Table 1: Major developing country and Annex I five biggest Carbon dioxide emitters as of 2005

SN	Country	Carbon Emissions (Million tonnes)	Carbon Emissions Per Capita (Million tonnes / Carbon dioxide/ Population)	Gross National Income (GNI) per Capita
	USA	5,816.96	19.61	43740
	China	5,059.87	3.88	1740
	Japan	1,214.19	9.50	38980
	Germany	813.48	9.87	34580
	Canada	548.59	17.00	32600
	United Kingdom	529.89	8.80	37600
	India	1,147.46	1.05	720
	South Korea	448.92	9.30	15830
	Mexico	389.42	3.70	7310
	Indonesia	340.98	1.55	1280
	South Africa	330.34	7.04	4960
	Brazil	329.28	1.77	3460
	World Totals	27,136.00	4.22	6,987

Source: *IEA, 2007a:48–57; World Bank, 2007: 288–289.*

1.2 Key concepts in the context of climate change

1.2.1 Climate

This is defined as the "average" weather conditions for a given place or region; it describes typical weather conditions for a given area based on long-term averages usually depicting rainfall, temperature, wind and humidity rates over the course of 30 years (or longer). However, some scientists think that it takes more than "average" weather to represent an area's climatic characteristics accurately (FAO[4] 2007). Although an area's climate is always changing, the changes do not usually occur on a time scale that is immediately obvious to human beings. We can observe how weather changes from day to day but climate changes are not as readily detectable (ibid). Hence, climate change and variability are gradual processes that occur beyond several generations.

A change of climate is considered to take place through the deviation of weather elements, the most important of which are: air temperature and humidity, type and amount of cloud coverage and precipitation, air pressure, wind speed and direction. A change in one weather element can produce changes in regional climate. For example, if the average regional temperature increases significantly, it can affect the amount of cloud coverage as well as the type and amount of precipitation that occur. If these variations occur over long periods, the climatic condition of an area will be regarded as changing (FAO 2007).

1.2.2 Climate variability

This refers to the short term deviation of climatic parameters of a region that are varying from its long-term mean due to internal processes such as vulcanicity and earthquakes, and external forces such as industrialisation, agricultural activities and urbanisation (FAO 2007). It is recognised that every year in a specific time period, the climate of a given location can be different with regard to temporal and spatial scales (FAO, 2007 & 2008). Some years may have below average rainfall, while some may happen to have average or above average rainfall. These changes result from atmospheric and oceanic circulation caused mostly by differential heating of the earth by the sun. Eventually, these atmosphere-ocean circulations cause climate to vary from one season to another or from year-to-year (FAO 2007).

1.2.3 Climate change

This can be conceptualised as the permanent deviation in weather conditions of a given area over an extended period due to both natural variability and

---

4 FAO – Food and Agriculture Organisation

anthropogenic processes (IDRC/CCAA[5] 2009; FAO 2007 2008; Paavola 2003). However, the IPCC (2001) definition of climate change is restricted to *climate change induced by human activities* (IDRC/CCAA, 2009). Climate change is manifested in the changes in the traditional patterns of every day weather and climate including extreme events such as high/low temperatures, droughts, hailstorms and floods (IDRC/CCAA, 2009). Variation in climate parameters is generally attributed to natural causes.

However, because of changes in the earth's climate since the pre-industrial era, some of these changes are now considered attributable to human activities (FAO 2007). Hence, climate change refers to any change in climate over time, whether due to natural variability or human activities (ibid). These natural and human processes are constantly altering the extremes of temperature, rainfall and air movement, which occur as natural processes. For example, droughts, periods of unusual dryness and excessive precipitation resulting in
floods are considered natural climatic occurrences. They are also regarded as unpredictable and unusual in that they do not occur all the time or occur only rarely in some areas (FAO 2007).

1.3 Climate change in Africa

For Africa, there has been a warming of approximately 0.7°C across most of the continent in the 20th century. Although these warming trends seem to be the same over the continent, climate changes are not always uniform. For instance, there have been decadal warming rates of 0.29°C in the tropical forests (Malhi & Wright 2004) and 0.1°C to 0.3°C in South Africa (Kruger & Shongwe 2004). In the same respect, in South Africa and Ethiopia, minimum temperatures have increased slightly faster than maximum or mean temperatures and between 1961 and 2000, there was an increase in the number of warm spells over Southern and Western Africa, and a decrease in the number of extremely cold days (New *et al.* 2006). A rate of warming of about 0.05°C per decade in Southern Africa has been observed during the present century (Hulme 1996; Jain 2006). The six warmest years in recent decades in Southern Africa have all occurred since 1980. The sub-region is expected to experience a mean temperature rise of 1.5°C and increased rainfall variability and insecurity (Hulme 1996).

Even more disturbing is the evidence of a looming average global temperature increase with warm temperatures ranging from 0.2 to 0.5 °C per decade. This warming is greatest over the interior of semi-arid margins of the Sahara and central Southern Africa (IPCC 2001). Effects of Global warming will also likely be seen on Mt Kilimanjaro as indicated in Photo 1. However, decreasing trends in

5 IDRC/CCAA –International Development Research Centre/Climate Change Adaptation in Africa.

temperature from weather stations located close to the coastal areas or to major inland lakes have been observed in Eastern Africa (King'uyu *et al.* 2000). This points towards the implication that there is need to consider that climate changes vary by location rather than blanketing changes as being the same over the continent. For instance, there may be a global increase in average temperatures but impacts of this global warming vary at lower scale – country and region.

Photo 1: Scientists have concluded that all snow glaciers on Mt. Kilimanjaro will be gone by the year 2050 (Observe retreating glacier line)

While inter-annual climate variability has been observed over most of Africa, multi-decadal climate variability in some regions has also been observed. On the one hand, observations in Western Africa indicate there has been climate change in the form of a decrease in mean annual rainfall since the end of the 1960s. For instance, between 1931 and 1960 and 1968 and 1990, there was a decrease of 30 to 40% (Chappell & Agnew, 2004; Dai et al. 2004; Nicholson et al. 2000). In the same respect, there was a decrease in mean annual rainfall in the tropical rain forest zone of 4% in West Africa, 3% in North Congo and 2% in South Congo for the period 1960 to 1999 (see Malhi & Wright 2004). On the other hand, there has been a significant increase in rainfall along the Guinean Coast in the last 30 years. While there has been a decrease in rainfall across most parts of the Sahel, an increase in rainfall has been registered in East and Central Africa (IPCC 2001). On the decadal time scale, Eastern Africa has been experiencing an increase in rainfall over the Northern sector and a decrease in rainfall over the Southern sector (Schreck & Semazzi, 2004).

In contrast, there are regions such as Southern Africa where no long term trends in climate change, especially in rainfall, have been noted (Boko *et al.* 2007; Richard *et al.* 2001). In some cases though, data inadequacies may mean that it cannot be determined if there have been changes (IPCC 2007). Instead, inter-annual variability is what has been observed in the post-1970 period. This variability has manifested itself in higher rainfall anomalies and the more intense and widespread droughts that have been reported (e.g., Boko *et al.* 2007; Fauchereau *et al.* 2003; Richard *et al.* 2001); including evidence of changes in seasonality and weather extremes (Boko *et al* 2007; New *et al.* 2006; Tadross *et al.* 2005a) a significant increase in heavy rainfall events has been observed in some parts of Southern Africa that include Angola, Zambia, Namibia, Mozambique and Malawi (Boko *et al.* 2007; Usman & Reason 2004). According to simulation studies, Southern Africa's precipitation will decrease by 5-20% in all major river basins of the region except the Congo where precipitation is expected to increase by 10% (Chigwada, 2004).

Photo 2: How much does tobacco farming contribute to climate change and global warming? Time to research on pros and cons.

The rainfall in the southern African region has been decreasing in the last 25 years but a lack of long-term trends in climate changes in Southern Africa implies that there could be less of climate change and more of climate variability in some areas in this region (Hulme 1996). However, it is important to acknowledge that adaptation in climate variability scenarios can be and has been used as a proxy for adaptation to climate change (Parry *et al.* 1999). A community's coping and adaptive capacities in the face of climatic variability and extremes are used as proxy for its level of coping and adaptive capacity for future climate change. Similarly, in the Third Assessment Report (TAR) of the IPCC (2001), it is argued that experience with adaptation to climate variability and extremes can be drawn upon to develop appropriate strategies for adapting to anticipated climate change (Parry *et al.* 1999; Usman & Reason 2004).

1.4 Climate Change Impacts on Africa

The impacts of climate change are likely to severely damage social and economic systems of most developing countries (IPCC 2001a; IISD 2008). More important, climate change implications for Africa are highly distinctive (Collier *et al.* 2008). Africa's climate has been largely characterised by droughts and floods and there are projections indicating that Africa is likely to be affected more severely than other regions. This is attributed to the continent's heavy dependence on agriculture and limited capacity to adapt (ibid). However, direct effects will vary widely across the continent. While some parts of the continent, such as Eastern Africa are predicted to get wetter, much of Southern Africa will likely get drier and hotter. Crop yields will be adversely affected and the frequency of extreme weather events will increase (Collier *et al.* 2008; Fluet *et al.* 2009).

Furthermore, the enormous effects of climate change on the African continent are compounded by the greater vulnerability of its economy (e.g. tourism, crop production and livestock keeping) to climatic variation (Collier *et al.* 2008). This has been considered to be unfair since Africa's past economic activity has not significantly contributed to the accumulated global stock of carbon, with its current activity accounting for only a trivial proportion of global emissions. Future projections suggest that Africa will continue to be the most affected by climate change impacts (Collier *et al.* 2008; Deressa & Hassan 2009). For this reason, while in developed countries the key issues are concerned with how to reduce carbon emissions, in Africa they deal with the adaptation of production to changing and mostly deteriorating climate conditions and subsequently, economic opportunities. In addition, while for these developed countries the main adverse consequences of global warming will be more apparent in the future and are still uncertain, in Africa many of the adverse consequences are already apparent (Collier *et al.* 2008). In this respect, it is also important for developing countries to consider the role that production of certain crops plays in global warming (see photo 2). The devastating effects of decreased rainfall are illustrated in Photos 3 and 4

Given this background, the key question every country must consider is how will climatic changes continue to affect African economies? By far the most important effects will likely be in agriculture, although these effects are also the least certain (Collier *et al.* 2008). The IPCC (2007) also concludes that current conditions of chronic hunger are likely to be made worse. This is because the proportion of arid and semi-arid lands is expected to increase by 5–8% by the 2080s and partly because of depleted water resources (ibid). However, the impacts of climate change on agricultural output will vary widely from country to country, with the IPCC (2007) projecting reductions in yield in some countries of as much as 50% by 2020. Small-scale farmers are expected to be the most vulnerable (Collier *et al.* 2008).

The vulnerability of small-scale farmers is coupled with the higher temperatures that have been directly changing crop yields (Challinor *et al.* 2007; Gregory *et al.* 2009). The area suitable for agriculture, the length of growing seasons and yield potential, particularly along the margins of semi-arid and arid areas are rapidly decreasing (Challinor *et al.* 2007). Gradually, temperature increase has triggered many crops in Africa to be grown close to their limits of thermal tolerance. It is known that just a few days of high temperature near flowering can seriously affect yields of crops such as wheat, fruit trees, groundnut, and soybean (Challinor *et al.* 2006). Such extreme weather is likely to become more frequent with global warming, creating high annual variability in crop production (Collier *et al.* 2008).

Furthermore, prolonged high temperatures and periods of drought will likely force large regions of marginal agriculture out of production. The maize crop over most of Southern Africa already experiences drought stress on an annual basis. This is likely to get worse with climate change and extend further southwards, perhaps making maize production in many parts of Zimbabwe and South Africa very difficult, if not impossible. Wheat yields in North Africa are also likely to be threatened (Collier *et al.* 2008).

In addition, it is proclaimed that environmental change induced by climate change is likely to inflict harsh and extreme environmental conditions upon rural smallholder farmers. These changes will likely have direct implications for creating sustainable livelihoods and may reduce the livelihood options of

Photo 3: Drought periods expose soil to other agents of erosion, especially wind.

poor farm households, especially within the agricultural and livestock sectors (Brown & Crawford 2008). A significant inter-relationship arises from the fact that insufficient water for livestock and crop production constitutes a main constraint to food security for millions of smallholder farmers, particularly in the semi-arid and sub-humid Sub-Saharan Africa (Assan 2008; Assan *et al.* 2009; Enfors & Gordon 2008; Falkenmark & Rockstrom 2004; Hahn *et al.* 2008 IPCC 2007a). By 2000, about 300 million Africans risked living in a water-scarce[6] environment. Moreover, by 2025, the number of countries experiencing water stress will rise to 18—affecting 75-250 million and 350-600 million people in 2020s and 2050s respectively (Arnell 2004; Hahn *et al.* 2008; IPCC 2007a). Such a scenario will likely exacerbate existing patterns of poverty and undermine policy attempts towards poverty alleviation, promoting sustainability and improvement in household wellbeing (Arnell 2004; Assan *et al.* 2009; Hahn *et al.* 2008; IPCC 2007a; Falkenmark & Rockstrom 2004).

However, attempting to understand the effects of climate change on Africa is impeded by several difficulties. While some facts regarding climate change are known and relatively well understood, there is still great uncertainty about

---

6 Water scarcity refers to the change of run-off regimes and the change (mostly lowering) of the groundwater table (UNESCO 2003a). Water stress on the other hand refers to a situation where water use exceeds water supply by 10% (UNEP 2002a).

Photo 4: Falling precipitation in the tropics: Rivers that had water flowing throughout the year have now turned seasonal, having water only during the rain season.

the key climatic processes. There is also much that is simply unknown (Collier et al. 2008). For instance, the African climate is determined at the macro-level by three major global drivers (the Inter-Tropical Convergence Zone (ITCZ), the El Niño–Southern Oscillation (ENSO), and the West African Monsoon), but how they interact and how they are affected by climate change is poorly understood (Collier et al. 2008).

What is probably known is that global warming affects the outcomes of key drivers of African climatic conditions; the ITCZ, ENSO and West African Monsoon; and thus increasing the incidence and severity of the droughts, floods and other extreme weather events that they produce (Collier *et al.* 2008). For example, the risk of drought in Southern Africa is related to the occurrence of the El Niño phenomenon in the Pacific, and there has been a tendency for these droughts to become more prolonged and frequent. In general, the drier subtropical regions will warm more than the moister tropics. Northern and Southern Africa will become much hotter (4°C or more) and drier (precipitation falling by 10–20% or more) (ibid).

Other simulations suggest that in Eastern Africa, including the Horn of Africa and parts of Central Africa, average rainfall is likely to increase by 15% or more (Collier *et al.* 2008). In general, the assumption is that many regions of

Africa will suffer from droughts and floods with greater frequency and intensity (Collier *et al.* 2008).

While the effects of climate change on agriculture are likely to predominate, there are three other effects of significance. High temperatures and higher peak temperatures will also affect health. High peak temperatures (above 30°C) will likely increase mortality, particularly in large conurbations. However, the effect will be modest in the overall context of African mortality (Collier *et al.* 2008). The effects via disease will probably be more substantial owing to the increase in disease-carrying insects. For example, the geographical distribution and the rates of development of mosquitoes are highly influenced by temperature, rainfall and humidity. We may expect an extension of the range of malaria-carrying mosquitoes and malaria into higher elevations, particularly above 1000 metres above mean sea level (Collier et al. 2008).

There have been resurgences of malaria in the highlands of East Africa in recent years. Although many factors are probably involved in the malaria implications, such as poor drug-treatment implementation, drug resistance, land-use change and various socio-demographic factors, including poverty, there is a strong correlation with climate change (Pascual *et al.* 2006 in Collier *et al.* 2008). This is attributed to fact that temperature in the highlands of East Africa has risen by 0.5°C since 1980, much faster than the global average. This is correlated with a sharp increase in mosquito populations (Collier *et al.* 2008).

Both malaria and dengue fever are expected to spread substantially unless countered. Malaria already inflicts enormous costs on Africa, over and above its direct effects on health. Gallup & Sachs (2001) in Collier *et al.* (2008) argue that controlling for other factors, the impact of intensive malaria is to reduce income by two-thirds; a 10% reduction in malaria is associated with 0.3% of higher growth. Hence the spread of malaria and the concomitant increase in the difficulty of its control may imply high though currently unquantifiable long-term costs (Collier *et al.* 2008).

In addition, the rise in global temperatures will in turn lead to a rise in sea levels by a metre or more by the end of the century. Such a rise will likely affect six million people residing in the Nile Delta. While island, beach and delta settlements seem to be dangerous in future, only a few parts of Sub-Saharan Africa are likely to be affected because most countries in the region do not yet have large urban population concentrations in the island, beach and delta settlements. However, the extreme low adaptation capacity, poor weather forecasts, poor disaster management and preparedness, poverty and high sub-division of Africa into many small countries implies that even these modest effects are highly

Photo 5: Persistent crop failures in rural areas render agriculture unsustainable leading to rural-urban migration, especially by the youth.

significant. For example, in Ghana the coastal zone occupies less than 7% of the land area but contains 25% of the population and so even relatively small rises could have damaging effects on the economy. Similarly, the tiny state of Gambia is at risk of having its capital city entirely submerged (Collier *et al.* 2008).

Greater exposure to flooding will have severe effects on infrastructure, most notably on the road system, which is currently predominantly unpaved and therefore particularly vulnerable to erosion from flooding. Although it is not possible to quantify the cost of the increased proneness to climatic shocks, typically, such a shock in a developing country reduces Gross Domestic Product (GDP) in the year of the shock by around 0.4% (Collier & Goderis 2008*a* in Collier *et al.* 2008).

Generally, developing countries are projected to experience impacts of climate change that stress their capacities to adapt before 2050 even at low climate sensitivity; at high climate sensitivity, some of these countries may be overwhelmed. Developed countries will become increasingly vulnerable. With high climate sensitivity, by 2100 much of the world may need not only a high adaptive capacity but also significant emissions mitigation to have been implemented in order to avoid high levels of vulnerability. Overall, these results challenge assumptions about which countries have "enough" adaptive capacity (because they are wealthy or impacts will be mild or both) (Yohe *et al.* 2006). Moreover, rural-urban migration will likely escalate as farming households livelihoods are threatened (see Photo 5).

Poorer developing countries are especially vulnerable to climate change because of their geographic exposure, low incomes and greater reliance on climate sensitive sectors. People exposed to the most severe climate-related hazards are often those least able to cope with the associated impacts, due to their limited adaptive capacity. This in turn poses multiple threats to economic growth, wider poverty reduction, and the achievement of the Millennium Development Goals (*Stern et al* 2006; Davies et al 2008). The social impacts, though generally not well-understood, are likely to be profound and climate change will affect humans through a variety of direct changes in climate variables and indirect pathways (pests and diseases; degradation of natural resources; food price and employment risks; displacement; conflicts, negative spirals) (Davies *et al.* 2008).

Planning for climate change in such situations will be extremely difficult when governments have limited authority and capacity to address the risks posed by existing hazards (Satterthwaite *et al.* 2007; Davies *et al.* 2008). Mainly due to climate change negatively impacting rural livelihoods; migration from rural to urban areas is increasingly likely to become the favoured adaptation strategy of the mobile, rural poor. This will further exacerbate the problem of people living in urban fringe hazardous environments with potential risks of social unrest (Davies *et al.* 2008).

1.5 Responsible Parties for Climate Change Mitigation and Adaptation

Signatories of the United Nations Framework Convention on Climate Change (UNFCCC) have committed themselves to addressing the "specific needs and special circumstances of developing country parties, especially those that are particularly vulnerable to the adverse effects of climate change" (Article 3) (Yohe *et al.* 2006).

The major developed countries emitting CO^2 are the USA, China, Japan, Germany, Canada and the United Kingdom. However, India is currently also contributing to greenhouse gases (IEA, 2007: 81; IISD 2008). While the major developing country emitters (e.g. India and Mexico among them) are poorer than developed countries, there may not be a large appetite in developed nations for wealth transfers to some of the major developing emitters, where rising economic power makes them global competitors whose influence is significant. These developing countries are competitive with developed countries on a range of products and services. The economies of China, India and many others are running fast and are fuelled by carbon intensive fossil fuels. This idea points to the need for developing country involvement, or at least the involvement of the major emitters among them, in a more meaningful way than was negotiated in the first commitment period. And this means finding a path forward that meets the needs of all parties, while accounting for the UNFCCC principle of common but differentiated responsibilities (IISD 2008).

CONCLUSIONS

This chapter has highlighted the importance of understanding the context of climate change in Africa before attempting to deal with these changes and variability. While there is a debate surrounding causes of climate change, there is more confidence to show that there is an upsurge of global warming leading to climate change. Evidence also shows that climate change is happening, is increasing and is largely accelerated by anthropogenic activities. Temperatures are highlighted as becoming warmer but with a reduction in precipitation. However, some parts of Africa such as those in the East have been observed to have incidences of increased rainfall, a factor which may be capitalised on for the livelihoods of rural communities. Moreover, the situation is gloomy for Africa as there is evidence to show that while this continent contributes the least to carbon emissions, it bears the brunt of their consequences more than countries in the developed world, which are largely responsible for these emissions. This is manifest in the impacts that Africa is experiencing from climate change.

CHAPTER TWO

FATE OF AFRICA IN A CHANGING CLIMATE
Climate Change Impacts, Vulnerability and Adaptation

2.0 Introduction

Chapter One has already highlighted the increasing evidence that climate change is occurring (Kristina & Johannes 2009). This occurrence can be traced back over the past two centuries; essentially from natural processes and human activities that have led to vast modifications and adverse consequences to the Earth's ecosystems and societies (IPCC 2007; Klein *et al.* 2005; Tschakert *et al.* 2008). However, the predicted consequences of environmental variability and climate change are diverse (Stern, 2007) and considerable uncertainty surrounds long-term patterns of environmental variability and their likely impacts on the livelihood activities and options of the poor (Brown & Crawford, 2008). Projections further suggest that by the end of the 21th century, environmental variability and change will have substantial impact on agricultural production and consequently, on the scope of reducing poverty in Sub-Saharan Africa, where the majority of the population resides in rural areas and depends on smallholder agriculture for their livelihood (Slater *et al.,* 2007; Assan *et al.,* 2009).

As the evidence for climate change impacts continues to mount, arguments about their consequences and implications remain a focus of local and international concern (Risbey, 2008; Blennow & Persson, 2009; IPCC, 2007; Grundmann, 2007). The concern arises from the ground that those impacts are evident through a complex mixture of positive and negative unevenly distributed impacts (see Section 1.4 for Climate change impacts on Africa). Extreme hazard problems are increasing beyond the ability of coping strategies to manage its effects (Collier *et al.* 2008).

Surprisingly, it is now more evident that climate change and variability bring significant negative impacts to countries that have historically contributed the least to greenhouse gas emissions, land-use change and have least capacity to adapt (Collier *et al.* 2008). For instance, the impact of climate change on Africa is likely to be severe because of adverse direct effects, high agricultural dependence, and limited capacity to adapt. However, direct effects vary widely across the continent, with some areas (e.g. Eastern Africa) predicted to get wetter, but much of southern Africa getting drier and hotter (ibid).

Given the pace and trend of climate change, the available literature recognises that climate change and variability, extreme events and structural changes have major impacts on economic, social and human living conditions as well as on natural systems. This implies, in developing countries, that key goals related to poverty reduction, water, food, energy, education and health are critically influenced by climate change and that adaptation measures should be tackled in the context of development policies specific to regions and communities, integrating local people's knowledge and adaptation skills (Halsnaes & Traerup 2009; Huq *et al.* 2003; Maarten *et al.* 2008). The negative impacts on water are cause for concern, given that water is a critical resource for the livelihood of communities (photo 6).

Whilst it is difficult to predict the current and future impacts of a phenomenon with such diverse impact streams, it is evident that the livelihood options of the poor and marginal societies in the global south are more likely to experience a significant shift. Again, their environment and the availability of the resources they depend on for their sustenance are predicted to dwindle (Assan & Kumar 2009). Therefore, policy approaches to find sustainable strategies and interventions have become more important to enable the continuity of natural

Photo 6: Availability of safe and clean water is increasingly becoming problematic with the changing climate in Africa.

and life supporting systems (ibid). Already, the WSSD[7] (2002) has provided a strong impetus to the discourse supporting links between climate policy and sustainable development and several UNFCCC[8] articles set out the provision for considering sustainable development (Klein *et al.* 2005). In developing countries, this concern is often fuelled by the fact that climate change has long-term trends and other repercussions on food security, water supply, sanitation, education and health care, all of which require more immediate attention (ibid).

The impetus provided by the WSSD has given rise to exploring and integration of policies and measures to address climate change in ongoing sectoral development planning and decision-making so as to ensure the long-term sustainability of investments as well as reduce the sensitivity of development activities to both today's and tomorrow's climate (Huq *et al.* 2003; Klein *et al.* 2005). Mainstream sustainable development has typically focussed on inter-generational equity issues and on global environment, particularly climate change and biodiversity depletion (Lumley & Armstrong 2004; Grist 2008). This has been taken into consideration because *the concept of sustainable development reflects development that meets the needs of the present without compromising the ability of future generations to meet their own needs* (WCED[9] 1987). It centres around three key areas of economic, social and environmental sustainability and the temporal aspect of continuity of the provision of material benefits (WCED 1987; Grist 2008).

2.1 Overview of Coping and Adaptation Strategies to Climate Change

The terms coping and adaptation reflect strategies for adjustments to changing climatic (environmental) conditions. In the case of a set of policy choices, both coping and adaptation denote forms of collective conduct that aim and indeed may achieve modifications in the ways in which society relates to nature and nature to society (Elsevier 2005). Coping strategies describe strategies employed during crises, whereas coping refers to the successes in dealing with a crisis. Adaptation is a characteristic of a favourable system or individual sector using coping strategies as part of standard behaviour. Adaptation strategies are then defined as a permanent change in which a household, community or ecosystem recovers from a crisis, irrespective of the time or the year in question (Dercon 2002).

---

7 WSSD World Summit on Sustainable Development that took place in Johannesburg - South Africa between August/September 2002.

8 UNFCCC United Nations Framework Convention for Climate Change.

9 WCED: World Commission on Environment and Development

Coping and adaptation strategies designed to reduce or recover from risks associated with changing climatic conditions also differ in their primary, immediate and long term strategies designed to either protect nature from society or society from nature. They have also differences in terms of significance to the concerned community or ecosystem (ibid). These terms are discussed separately in the following sections.

2.1.1 Coping Strategies

Coping strategies encompass both immediate responses (e.g. sources of off-farm income, post-disaster financing sources, and emigration plans) and more structural and long-term strategies, such as re-orientation of production and improvement of infrastructure for production. It also includes strengthening entitlement systems, community empowerment, livestock grazing on failed plots, asset sales for cereal purchases and asset distribution, food transfers and migration for employment and human–environmental relations such as changes in land use and water management strategies, among others (Saldana-Zorrilla 2008; Enfors & Gordon 2008). Several coping mechanisms have been revealed throughout Africa. For instance, migration and environmental change

Photo 7: Apart from farming activities, peasants in rural areas also engage in other activities such as petty-trading to supplement their harvests.

have been documented as being acutely pertinent to sub-Saharan Africa (Unruh *et al.* 2005; Unruh 2008). This trend of migration, whether permanent or temporary, national, regional, or international, has always been a possible coping strategy for people facing environmental changes (Hamza *et al.* 2009). Petty trading has also become an import way of coping with climate variability as illustrated in Photo 7.

Several studies have indicated technical insights on how farmers respond to stress. These studies indicate that farmers may stagger sowing times, increase sowing densities or re-sow if germination appears to be low (McGuire 2007). In response to rapidly changing environmental conditions, labour supply or even market signals, farmers may also alter types of crops grown, relative crop areas or variety portfolios (McGuire & Sperling 2008). Farmers also respond to stress by making proportionally greater use of off-farm channels as harvests tumble or seed quality declines. In the same vein, markets and social networks may be drawn upon to fill shortfalls or to help farmers switch toward specific crops or varieties (Sperling *et al.* 2008). However, sometimes access to channels is often compromised post-crisis by farmers' low purchasing power or weakened functioning of social networks, such as community total seed shortfalls (McGuire 2008), or because neighbourly sharing has broken down (as sometimes occurs following conflicts). Therefore, ensuring community resilience to climate change depends on its infrastructure, neighbourhood relations and internal functioning of socio-economic systems and institutions (McGuire & Sperling 2008).

2.1.2 Adaptation Strategies

Adaptation to climate change refers to adjustment in natural or human systems in response to actual or expected climatic stimuli or their effects. These strategies are employed to moderate harm and exploit beneficial opportunities (IPCC 2001, 2007; Carr 2008; Grist 2008; Deressa *et al.* 2009). Adaptation to climate change takes place through adjustments to reduce vulnerability or to enhance resilience. Hence, adaptation includes taking action to reduce risk or enhancing benefiting from available opportunity (Blennow & Persson 2009).

Some of the adaptation strategies used include the use of new crop varieties and livestock species that are better suited to drier conditions, crop diversification and the use of different scales of irrigation systems, changes in fertilisation techniques, reduced utilisation of marginal lands, water harvesting practices and water resources conservation as well as precision farming methods (Bradshaw *et al.* 2004; Kurukulasuriya & Mendelsohn, 2008; Nhemachena & Hassan 2007; Halsnaes & Traerup 2009; Deressa *et al.* 2009).

These technologies transform a fully rain-fed to a fully irrigated continuum and improve plant-available water by bridging dry spells, minimizing run-off, drainage and evaporation and maximizing infiltration and soil water holding capacity (Deressa *et al.* 2009). In addition, potential adaptation measures against flooding include construction of roads with drainage systems, increased road levels, alternative routes, stronger foundations and bridges (Halsnaes & Traerup 2009). Field experience has shown that adaptation to climate change in Southern Africa includes crop diversification, different planting dates, planting different varieties, increased use of crop varieties, use (or increased use) of irrigation or water and use of water and soil conservation techniques (Grist 2008).

Furthermore, it is important to note that adaptation can be attained in two forms (Davies *et al.* 2009); reactive and proactive. On the one hand, *reactive* adaptation focuses on *coping* with the adverse impacts of climate change when they occur. On the other hand, *proactive* (or anticipative) adaptation encompasses measures taken in advance to limit the ultimate damages of climate change or to reduce the extent of reactive adaptation required when climate change impacts materialize. For example, evacuating people from a flood-hit area is reactive adaptation, while modifying zoning laws on coasts in anticipation of stronger sea surges is proactive adaptation (ibid). However, it is important to note that the distinction between proactive adaptation and reactive adaptation is intuitively clear but difficult to delineate with precision in a dynamic setting (ibid) since both approaches can bring immediate and long-term measures to the changing climate.

2.2 Synergistic approaches for Coping and Adaptation Strategies

Climate change may not turn out to be the greatest challenge humanity faces in the twenty-first century, but there is no doubt it requires extremely serious and sustained global attention (Hepburn & Stern 2008). Increased understanding of climate threats to the natural environment and human livelihood systems such as agriculture is therefore fundamental. Africa in particular has not been met with an improved understanding of how best to respond to changes in climate (Burke *et al.* 2009). It is also proclaimed that problems from climate change will arise especially for places that are not used to dealing with particular hazards or are accustomed to lower levels of intensity (Maarten *et al.* 2008).

Therefore, the dangerous impact of the hazards will fall especially on vulnerable people who do not have experience with these hazards (Maarten *et al.* 2008). For such trends, immediate and long term measures should be central to any development initiative, particularly through informing local communities about the ramifications of climate change and integrate traditional knowledge to generate solutions likely to work at their level (Maarten *et al.* 2008). Such consideration should focus on assessing communities' perceptions of changing conditions, as well as the way coping strategies are being applied, stretched, and possibly modified (Maarten *et al.* 2008).

Given the adaptability displayed by farmers in the face of past climate variability (Adger *et al.* 2005), African farmers are likely to continue coping and adapting to climate changes, for example by adopting new crops or varieties or by altering the timing of planting and other agronomic practices. But if future climates move as quickly outside the range of past experience as they are expected to throughout the tropics, farmers may be unable to adapt rapidly enough without some help. As a result, there exists a widely acknowledged need for significant investment in agricultural adaptation. However, this strategy can be constrained by a little systematic assessment of how to prioritize planned adaptations; especially what form they should take, and on what crops, environmental resources and locations they should focus (Burke *et al.* 2009).

For instance, one form of adaptation that has moved very slowly in Africa is technology adoption. It is widely noted that Africa lags behind the rest of the world in adopting large-scale irrigation, capital, and high-yield varieties (Evenson & Gollin 2003). However, adoption of these technologies may help farmers adapt by increasing productivity and counterbalance losses from climate change, particularly in drier or hotter conditions. Water harvesting for small scale irrigation may enable farmers to deal with challenges of unavailability of water for agriculture (Photo 8). A case in point is the typical development of new soybean varieties in Brazil. Thus, this weakness can be addressed through research and outreach; facilitating the development and use of crop varieties with more tolerance for the hot and dry conditions of many of Africa's agro-climatic zones (Kurukulasuriya *et al.* 2006).

While climate change has become a global challenge for present and future environmental and livelihood concern; coping strategies vary widely within communities. They are established in the context of specific local settings and along with the capacities and livelihood strategies of individual actors (McGuire & Sperling 2008; Thornton *et al.* 2009). Also, while coping with climate change and variability is not a new experience for African farmers, existing coping mechanisms may not be up to the challenges posed by the changes projected. The situation is made more complex by the fact that while we know something about the changes possible in climate in future years, we know much less about likely changes in climate variability and the probabilities of extreme events. This is particularly expected to affect vulnerable people who are highly dependent on natural resources for their livelihoods (IPCC 2007; Thornton *et al.* 20

Photo 8: Irrigation agriculture is an important alternative to rain-fed agriculture.

In addition, although households have considerable experience of coping with temporary shocks, defensive flexibility has not been combined with sustained ability to adapt to new circumstances or adopt new technologies. One manifestation of this is Africa's continued dependence upon the same narrow range of commodity exports. At the microeconomic level, technical progress has been slower than in other regions, both among farms and manufacturing firms. Thus, it might be more important to consider the integration of coping measures into long term adaptation enhancement (Collier et al. 2008).

While climate impacts seem to threaten agricultural development, small-scale water system technologies appear to have the potential to stabilize and increase yields in current farming systems of vulnerable communities in most developing countries (Ngigi 2003; Makurira *et al.* 2007b). Hence, for effective approaches towards integrating coping measures into adaptation scales, small scale irrigation systems that rely on traditional approaches may be transformed into large-scale operations as a way to upgrade rain-fed agricultural productivity in semi-arid and dry sub-humid sub-Saharan Africa (SEI 2005; Enfors & Gordon 2008).

Furthermore, breeding crop varieties to tolerate future climates has contributed to unprecedented gains in human well-being in some regions such as America and Asia (Evenson & Gollin 2003 in Burke *et al.* 2009). Therefore, the development of improved crop varieties suited to Africa's diverse agro-ecologies can facilitate agricultural improvements. While advances in biotechnology can offer alternatives, more direct pathways to genetic modification of crops, traditional breeding has a long history of success in many parts of the world and is typical for most African breeding programmes and thus will likely continue to be a critical strategy for crop improvement by producing cultivars resistant to drought, heat or a particular pest or disease (Burke *et al.* 2009).

Generally, coping and adaptation measures need to be made at policy level to reflect short-term, medium-term and long-term goals. Short-term goals are likely to include measures to address food security, clean water, local employment, and stable disaster management frameworks. In the medium-term, the development of forecasting mechanisms may improve coping and adaptation to climate change and the preparedness for extreme hydrological events (Chhibber & Laajaj 2008). Long-term goals such as economically viable development pathways, equitable social structures and successful system transformability, on the other hand, may provide small scale and large scale design on building community resilience to climate change impacts (Tschakert P*etal.* 2008).

Other suggestions for effective adaptation include strategies that can involve capacity building, the enhancement of the skills of people and the capacity of institutions in resource management through education and training (Wescott 2002). This can be done hand-in-hand with the creation of appropriate policies and institutions in order to cope and adapt effectively. Capacity building should be a multi-sectoral, long-term process and the transfer of this knowledge must be done in a manner which will ensure longevity and sustainability of coping and adaptation measures (Crabbe 2009).

Moreover, it is important to undertake multi-sectoral assessment to investigate the strength of people's current resilience and capacity to adapt. Such assessment of coping and adaptation capacities typically must involve a range of coping measures (for instance, access to extended networks of mutual assistance and other forms of social capital, cropping adaptations informed by local knowledge of climate indicators, adjustments to expected slow-onset floods and drought preparedness especially in semi-arid areas) (Maarten *et al.* 2008).

Although these strategies may not be sufficient to cope with the new challenges brought by climate change (Maarten *et al.* 2008), they may provide a framework for better-informed choices and priorities for future adjustment to the climate change impacts (ibid). Such information must base on actual experience and in perceived local priorities and integrated to the prepared people to cope with conditions that they have not yet experienced. However, feeding new information about future trends must be done with caution in order to prevent people from focusing only on that new information (ibid).

Generally, successful coping and adaptation measures to climate change challenges should involve both public and private-sector response; changes in the sectoral structure of production and changes in crop patterns. The role of government should also focus on the provision of the information, incentives and economic environment to facilitate sustained coping and adaptation measures (Collier *et al.* 2008).

2.3 Ensuring Sustainable Development through Adaptation Mechanisms

As dynamics of natural and social transformation have become greater and speedier, so too have the opportunities for adaptation. The realization of multiple goals by means of adaptive strategies is conceivable. Adaptive processes can become the engine of sustainable management and development if there is expertise that links local understanding of disasters, health systems, water resources, agriculture and energy. Also, this can adequately be meaningful if climate change information is to be translated into practical parameters for decision-makers, policy-makers and the general public awareness (Maarten *et al.* 2008). As a result of such actions, it is ultimately possible to speak of a reduction of greenhouse gases by means of adaptation and multiple political, economic and legal strategies that may be deployed to ensure that the damages from climate change are lowered (Elsevier 2005).

Therefore, such challenges imposed by climate changes in natural and human systems require a need for establishing adaptation measures to promote sustainable development. It is possible that sustained natural and socio-economic systems emanate from proper adaptation as adaptive strategies take effect more rapidly and the benefits become evident much more promptly. Adaptive processes have a comparatively brief time horizon and the capacity of science and technology to innovate is more easily realized in adaptive measures. Again, adaptation is possible without special incentives and it is characterized by a lower ambiguity concerning the kinds of tasks that are necessary and the ways in which their accomplishment can be assessed. Finally, knowledge about adaptation is more robust and can likely be assimilated more readily to what might be called forms of "practical knowledge" (Stehr 1992). The risks and dangers associated with uncertainty are fewer in the case of adaptive measures (Elsevier 2005; Luhmann 2005).

In addition, effective adaptation has to operate within the scope and capacity of personal action. Hence, adaptation must not be regarded as something done to or for people; it must be something that they do for themselves and that may (or may not) be supported by external agencies (IISD *et al.* 2003 p. 16). In this vein, the purpose of adaptation activities is thus to: 'sustain existing and open up new livelihood opportunities and to help forge stronger and more cohesive community-level institutions whilst mainstreaming it into wider development (IISD *et al.* 2003 p. 16). The aim is to seek 'win–win' solutions that strengthen immediate needs provision 'and also contribute to longer term capacity building and structural change' (IISD *et al.* 2003 p. 18). What this structural change might be and how radical is difficult to imagine within this perception of adaptation as the process of people responding to the effects of climate change 'for themselves' by improving their short- and long-term livelihoods' resilience (Grist 2008).

Also, promoting sustained development through adaptation measures requires the integration of the diversity of sectoral actors and the interconnectedness of climate change and development policies at local and global levels. Specifically, policy must address four roles. These will involve: controlling the atmospheric concentrations of greenhouse gases; preparing for and reducing the adverse impacts of climate change; and, taking advantage of opportunities and addressing development and equity issues. Although climate change is not the primary reason for poverty and inequality, addressing these concerns is seen as a prerequisite for successful climate policy in many developing countries (Klein *et al.* 2005). A fourth role of climate policy will reflect facilitating the successful integration and implementation of mitigation and adaptation in sectoral and development policies. For climate policy to take on this role effectively and efficiently, research is required to provide data sets for proper implementation of policy objectives (Klein *et al.* 2005).

While private actors are naturally forward-looking and have undertaken gradual steps towards climate change adaptation, private actors can be presumed to respond appropriately to changing conditions only if they have adequate information, appropriate incentives and an economic environment conducive to investing in the required changes. Hence, the most promising strategy for government is to ensure that two conditions are met: providing incentive to act and providing the information and incentives those private actors need in order to induce adaptation. The most information-intensive aspects of adaptation are likely to be changes between crops and crop varieties (Collier *et al.* 2008).

International cooperation can assist adaptation, particularly the cooperation between governments within the region. It has been indicated that the shocks to any region as a whole are considerably smaller than those to individual countries. This suggests that there might potentially be benefits from regional integration and these could be derived through several different channels. The first is that regional integration might facilitate migration. Although this strategy may encounter local migration barriers, it can facilitate movement of labour force. The second channel is trade since countries may become more dependent on food trade, both because of changing patterns of comparative advantage and because of the need to integrate markets to pool risk. The third potential benefit of regional integration derives from improved cooperation on non-market inter-country linkages, in particular water (Collier *et al.* 2008).

Promoting irrigation can help alleviate the likely effects of climate change in Africa. Where water is available, moving from dryland to irrigation system agriculture can increase average net revenue per hectare and resilience of agriculture to climate change. Therefore, governments can make public investments in infrastructure and canals for water storage and conveyance where appropriate and where the public good nature of these investments prevents adequate private sector investment (Kurukulasuriya *et al.* 2006).

Finally, in addition to encouraging direct adaptations, both local and national governments and international organizations can invest in infrastructure and institutions to ensure a stable environment to enable agriculture to prosper. Such policy interventions may achieve the long-term goal of helping vulnerable populations adapt to climate change and increasing the likelihood of achieving the more immediate Millennium Development Goals, such as halving hunger, reducing poverty and improving health (Kurukulasuriya *et al.* 2006). Generally, developing country populations with their higher vulnerability to climate change, lower resilience and limited adaptive capacity are conceived as a key locus of concern (IISD *et al.* 2003; Poverty & Environment Partnership 2003). Adaptation has become a major theme in climate change and development (Eriksen *et al.* 2005; Huq *et al.* 2006; Hulme *et al.* 2001).

2.4 Challenges across Spatial and Temporal Scales

It has already been (see Chapter One) acknowledged that climate change has implications for Africa which are highly distinctive and Africa's climate is likely to be affected more severely than that of other regions (Collier *et al.* 2008; Deressa & Hassan 2009; Thornton *et al.* 2009). This is compounded by the far greater exposure of Africa's economy to climatic variation (Collier *et al.* 2008; Enfors & Gordon 2008). In contrast to this atypically severe exposure to effects on production, Africa's role in emissions of carbon is atypically minor (Collier *et al.* 2008; Deressa & Hassan 2009).

Despite adaptation measure in place across regions, the current global socio-economic measures are characterised by the promotion of win-win situations. In many cases win-win outcomes cannot be feasible and there will be winners and losers, particularly vulnerable communities in rural areas of developing countries (Unruh 2008). For instance, semi-arid and dry sub-humid Sub-Saharan Africa (SSA) presents large challenges in terms of achieving the Millennium Development Goals on eradicating poverty and hunger (Burke *et al.* 2009; Enfors & Gordon 2008). It is estimated that 45–50% of the population in this region live in extreme poverty and the level of malnutrition is high (UNDP 2006). The main livelihood source is small-scale rainfed farming but yield levels in current production systems are very low (Enfors & Gordon 2008; Burke *et al.* 2009).

Ultimately, as a response to the low yielding farming and as a way to accumulate wealth, income diversification is becoming increasingly common (Ellis 1998 in Enfors & Gordon 2008). Livelihood security for smallholders in these regions is still intimately linked with the local agro-ecological productivity, which is largely constrained by water availability. Although the cultivated dry lands in SSA are characterized by reasonably good seasonal rainfall, the distribution is extremely unpredictable, making the capacity for coping with temporal water shortages essential for farmers (Falkenmark & Rockstrom 2004). Common strategies among smallholders to handle rainfall uncertainty include crop variety diversification, matching labour inputs to expectations of the season, livestock grazing on failed plots, asset sales for cereal purchases, food transfers and migration employment (Enfors & Gordon 2008).

Despite the existence of many studies highlighting farmers' strategies for coping with stress (e.g., Mortimore & Adams 2001; Thornton *et al.* 2007; McGuire & Sperling 2008), it is rare that interventions addressing vulnerability engage with or build upon these strategies (McGuire & Sperling 2008). For instance, most efforts promoting adaptation to climate change overlook the adaptive capacity of vulnerable populations (Reid & Vogel 2006). This effectively treats farmers as passive victims, denying their agency in responding to hazards (Tschakert 2007). Approaches to reduce farmer vulnerability tend to be supply-driven as they reflect what interventions are on hand rather than the needs arising from a specific local setting (the demand side). Such a gap leads to 'one size fits all' interventions, which are not necessarily effective at reducing farmers' vulnerability and may even have inequitable outcomes (Eriksen *et al.* 2005; Adger *et al.* 2003; McGuire & Sperling 2008). However, there is growing recognition that efforts to strengthen the resilience of systems need to understand and build upon local coping strategies (McGuire & Sperling 2008).

Furthermore, an appropriate response to changes in the relative productivities of location is the relocation of labour and capital towards the relatively favoured places. However, conditions of some regions, Africa in particular, present several impediments. The movement of people is constrained both by the informal restrictions of strong ethnic identities and the formal restrictions of national boundaries. Africa is even more sub-divided into ethnic groups than it is into countries so that ethnic identities create barriers to movement even within a country. There is currently controversy over the extent to which population movements owing to the drought in the Sahel have contributed to the conflict in Darfur as pastoralists seeking water resources clash with sedentary arable farmers. Cross-border migration on a mass scale can be stymied by political restrictions, and even where it is permitted has the potential for violent conflict. For example, the movement of populations from arid and landlocked Burkina Faso to coastal Côte d'Ivoire at its peak was so large that around 40% of Ivoirian residents were Burkinabe. This facilitated populist politics that were instrumental in triggering a political collapse into civil war (Collier *et al.* 2008).

Other scholars have proposed institutional mechanisms to facilitate proper movement of 'climate change exiles' or 'environmental refugees' as a form of compensation for climate change impacts (Byravan & Rajan 2005, 2006; Myers 2002). Arguing against these proposals, Adger & Barnett (2005) suggest that encouraging migration as a solution to climate change detracts from the need for adaptation policies to allow people to 'lead the kind of lives they value in the places where they belong' (Mortreux & Barnett 2009). At least two key issues need to be considered in weighing up whether an individual may migrate due to climate change: what they perceive to be the risks associated with climate change and how they analyse the benefits and costs arising from migrating/staying. It is the way these spheres intersect that determines possible migration responses to climate change and the attainment of sustainable development (ibid).

Even where people are able to migrate across ethnic or national boundaries, they may not be able to gain access to land. In most Africa land rights still reflect some ancestral claim and are not readily marketable. The reduced productivity of many locations is largely specific to agriculture, the sector which accounts for more than 60% of the African labour force. Therefore, this can hamper technological diffusion among farmers in major parts of Africa. Further, the adverse effect of climate change on agriculture is much more pronounced in Africa than in other regions. The climatic variation will be more severe in Africa and African agriculture is still overwhelmingly rain-fed and so more vulnerable. This has two important implications, one macroeconomic and the other concerning comparative advantage and the inter-sectoral allocation of resources (Collier *et al.* 2008)

In respect of one key public good, the road network, climate change constitutes technical regress (see Photo 9). Increased climatic variation and in particular, intense bouts of rainfall can dramatically erode unpaved surfaces. The appropriate response here is to an extent ambiguous. Overall, the return on investment in roads is lower and so the appropriate size of the road network is smaller than would otherwise be the case. However, because the cost of maintaining a road network of any given size increases, the appropriate level

Photo 9: Poor infrastructure in rural areas is another barrier to enhancement of people's capacity to effectively cope with changing climate.

of expenditure on roads may well rise. This is just one example of how the provision of public infrastructure will have to be adapted to climate change. As it becomes clearer the form climate change will take in particular countries, so action will have to be taken to 'climate-proof' infrastructure investments (Collier *et al.* 2008).

The record of African government investment in large-scale irrigation suggests that there are major difficulties with this approach, although there may be more scope for financing smaller-scale and local-level cooperation. Potentially, the government could require change by means of effective regulation. However, the history of agricultural regulation in Africa suggests that usually the regulatory route would meet such strong resistance as to have only limited effect. Actually, most African governments lack the administrative capacity to enforce complex agricultural changes that are against the private interests of farmers (Collier *et al.* 2008).

As an additional threat to African development, access to water resources (Photo 10) is likely to be the source of an increasing number of conflicts in the future (Stern 2007). National as well as cross-border conflicts motivated by water access have been observed already. In Mali, the 1970s and 1980s droughts forced many semi-nomadic Tuareg to migrate; their troublesome return to their native lands was the basis for the 'Second Tuareg Rebellion' in 1990. The use of dams along the Senegal River provoked clashes between Senegalese and Mauritanian populations during the late 1980s and early 1990s (Niasse 2005). The West African region has already experienced a decline in its rainfall by 10% to 30% during the past three decades. This raises a lot of concerns for the forthcoming decades. Hence, cooperative mechanisms will be required to prevent the commencement of additional water-related contentions (Chhibber & Laajaj 2008).

On addressing climate related diseases, there are several adaptation measures which can be applied, especially for malaria prevention. One option is residential house spraying which involves treating all interior walls and ceilings with an insecticide, which is effective against mosquitoes that favour indoor resting before or after feeding. However, spraying has been abandoned in many countries due to disillusionment over eradication results and to concerns over safety and environmental impacts. Administrative, managerial and financial constraints have also been an implementation barrier (Halsnaes & Traerup 2009).

Other options are malaria vaccine and medical prevention with chemoprophylaxis - a presumptive intermittent treatment. However, this is not perceived as appropriate to the whole population. Lastly, Insecticide Treated bed Nets (ITNs) is an option, which have shown to be an effective, feasible and economically attractive measure in reducing mortality and morbidity from malaria. ITNs effectively provide a reduction in transmission intensity as they prevent mosquito bites and shorten the mosquito's life span, thereby reducing transmission (Breman *et al.* 2006). The beneficial impacts of large-scale ITN programmes have been demonstrated in Tanzania and recently in Eritrea and expansion of ITN coverage is chosen as the adaptation measure in this example (Halsnaes & Traerup 2009). However, this measure lacks general public awareness about the use of nets.

Further, regional trade in food is proving to be so vulnerable to political intervention that it is unreliable. For example, in response to the rise in global food prices the government of Tanzania imposed restrictions on food exports to Kenya, despite both belonging to the East African Community (EAC). The third potential benefit of regional integration derives from improved cooperation on non-market inter-country linkages, in particular water. While increased rainfall in East Africa may increase flow in the Nile, drier and hotter conditions in Southern Africa may reduce the flow of the Zambezi significantly. There is potential for inefficient and inequitable water use and, ultimately, conflict may occur unless regional cooperation is established (Collier *et al.* 2008).

Photo 10: How much rural communities efficiently and effectively use water resources such as rivers is an area where further research is wanting.

Generally, adaptation to the impacts of climate change are not well organised at the global and regional scale. Some adaptation by individuals are planned while others may be spontaneous reactions to changing circumstances related to resource use (e.g. forestry and agriculture) or related changing economic constraints or opportunities. There is clearly a role for public policy in climate change adaptation in creating the environment for appropriate adaptation (Adger *et al.* 2001).

CONCLUSIONS

It has become evident that rural people will need to change their lifestyles to adapt, either because of two reasons, that local impacts of climate change leave them no alternative and also that specific adaptation will reduce the losses associated with those impacts substantially. Climate change events place communities at risk and climate change adaptation has increasingly been viewed as an issue of risk management. This has prompted considerable and increasing activity on the part of development agencies and governments to come to grips with these challenges, including the development of appropriate adaptation strategies. Given the scale of the problems involved, development agencies could greatly benefit from information that quantifies the impacts that may arise so that development assistance can be targeted in appropriate places, depending on the development objectives that are being pursued. Therefore, it is important for research to target the considerable knowledge gaps concerning the interacting and multiple stresses on the vulnerability of the poor in Africa. Analytical assessments of vulnerability to increased climatic variability and climate change to better understand the implications for poverty reduction as well as to be able to assess adaptation initiatives cannot be overemphasised.

CHAPTER THREE

COMPLIANCE TO FINANCIAL REQUIREMENTS FOR
ADAPTATION TO CLIMATE CHANGE IMPACTS

3.0 Introduction

Due to significant impacts on economic, natural and social systems, climate change is now commonly identified as one of the most urgent and critical issues for the global community to address. The Fourth Assessment Report (AR4) of the Intergovernmental Panel on Climate Change (IPCC 2007a) confirmed that the warming of the climate system is unequivocal and human actions are changing the earth's climate and creating major disturbances in human systems and ecosystems (IIED 2008). The international community has long recognized the need to support adaptation to climate change as indicated by its inclusion in the United Nations Framework Convention Climate Change (UNFCCC) (IISD 2008).

Articles 4.3 and 4.5 of the UNFCCC both call for developed countries to provide new and additional financial resources to meet the agreed costs of developing countries in complying with their obligations under the UNFCCC. This includes implementing measures to mitigate climate change by addressing anthropogenic emissions by sources such as fossil fuel combustion and removals by sinks (UN 1992). In addition, Article 11.5 stipulates that developing countries may avail themselves of financial resources related to the implementation of the Convention through bilateral, regional and other multilateral channels. More specifically, the Convention established a new financial mechanism, the Global Environment Facility (GEF), for the provision of financial resources on a grant or concessional basis. Also, under Article 4 of the Convention, developed countries are required to assist developing countries that are particularly vulnerable to the adverse effects of climate change in meeting costs of adaptation. No specific provisions are included in the Convention for financing mitigation and adaptation actions of developed countries; rather it is assumed that these countries have mechanisms to finance their own activities (IISD 2008).

Furthermore, four dedicated funds have been established under the current climate regime to support adaptation to climate change. These include the Least Developed Countries Fund (LDCF), which is replenished through voluntary contributions from Annex II countries. The fund was established to support the Least Developing Countries (LDCs) in the development of National

Adaptation Programmes of Action (NAPAs) and is beginning to finance priority projects identified under these Plans. Another funding mechanism is the Special Climate Change Fund (SCCF), which addresses adaptation and technology transfer along with emission reduction projects and economic diversification. Also, the Adaptation Fund was introduced under the auspices of the Kyoto Protocol to support "concrete adaptation projects and programmes" in developing countries that are Parties to the Protocol. The Adaptation Fund is funded by 2% levy imposed on transactions of certified emission reductions (CERs) from Clean Development Mechanism (CDM) projects. Finally, is the Strategic Priority on Adaptation (SPA), established by the GEF under its Trust Fund in 2004 at the request of COP-7. The goal of the SPA is to support pilot and demonstration adaptation projects that reduce local vulnerability to climate change and provided global environmental benefits (IIED 2008)

However, the general global attention on climate change is predominantly focused on carbon and methane emissions and mitigation. There are some actions that Africans can take to reduce their emissions, particularly to do with land use and deforestation, which are by far the most important aspects of mitigation for Africa. At one extreme, some strategies for global mitigation have serious adverse consequences for Africa and so would be damaging for the region even if they succeeded in arresting global warming. At the other extreme, some strategies create new income-earning opportunities for Africa, although whether these opportunities are harnessed is again contingent upon human response (Collier *et al.* 2008).

International actions to assist African adaptation have both an ethical and a practical rationale. The ethical rationale is evidently that the rising levels of carbon dioxide are not attributable to Africa but to economic activity elsewhere. Therefore, there is a strong case that the costs of African adaptation to these adverse externalities should be borne by others. In practical terms, some of the public goods required for adaptation are regional in nature, such as climate information, agricultural research and transport infrastructure. Africa's extreme political sub-division means that it faces intense problems in supplying regional public goods and international assistance can usefully substitute for these missing public goods. Assistance for adaptation also needs to be seen more broadly in the context of development policy (Collier *et al.* 2008).

There is acknowledgement that climate change is likely to make development more difficult and diversification out of agriculture more necessary. This suggests that OECD trade preferences for African non-agricultural exports might become more important (Collier *et al.* 2008). Aid flows might need to be reallocated between sectors to support such diversification and reallocated geographically to target those societies most severely affected. Finally, the most elementary step would be for the international community to do no harm.

3.1 Expected financial requirements

The Stern Review (2006) estimates that the costs of reducing GHG emissions to avoid the worst impacts of climate change can be limited to around 1% of global GDP each year. Without action, the overall costs and risks of climate change will be equivalent to losing at least 5% of global GDP each year and estimates of damage could rise to 20% of GDP or more if a wider range of risks and impacts is taken into account (IISD 2008). In the absence of mitigation and adaptation efforts, the economic damage caused by climate change will potentially be in the *trillions* of dollars per year. In the near term, a temperature rise of 2°C to 3°C (as is expected to take place within the next 50 years) is projected to result in a permanent economic loss of up to 3% of global GDP (Stern 2006). Planned adaptation measures can reduce these costs. The scale of investment required to undertake these measures, however, is highly uncertain. This uncertainty reflects existing limitations in our knowledge of the type, magnitude and timing of climatic changes and their subsequent impacts, as well as the long time horizons involved. However, some initial estimates provide an indication of the expected scale of financing that could be needed:

- The UNFCCC has estimated that in 2030, between US$49 and $171 billion dollars in additional investment and financial flows will be needed globally for adaptation; of this amount, US$28 to $67 billion will be needed by non-Annex I Parties (UNFCCC 2007a);

- The estimated additional cost of climate-proofing new infrastructure and buildings in OECD countries could be between US$15 and $150 billion per year (or 0.05 to 0.5 per cent of GDP; Stern 2006);

- The World Bank (2006a) has estimated that approximately 20 to 40% of activities financed by Official Development Assistance (ODA) and concessional finance are sensitive to climate risks and that the annual cost of addressing this risk would be US$1 to $8 billion;

- Additionally, the Bank has estimated that between US$9 and $41 billion will be needed annually to "climate proof" *new* investments globally (World Bank 2006a; IISD 2008);

- Oxfam has estimated that US$50 billion per year will be required each year to assist developing countries with their efforts to adapt to climate change (Oxfam 2007a); and

- The cost of adaptation priority activities identified in 16 of the first 17 National Adaptation Programmes of Action submitted by LDCs to the UNFCCC amounts to US$292 million (UNFCCC 2007a; IISD 2008).

Although these estimates are generally derived from basic calculations and make a number of assumptions, they suggest that tens of billions of dollars in additional funding will be required each year to help countries prepare for and respond to unavoidable impacts of climate change. These funds will need to be provided through a combination of national and local government expenditures (in developed and developing countries), private sector investments, and the transfer of funds from developed to developing countries (IISD 2008).

3.2 Technological Transfer

Under the changing climate, technology transfer has been placed as critical to achieving the goal of the United Nations Framework Convention on Climate Change (UNFCCC-Article 2) to achieve "stabilization of greenhouse gas concentrations in the atmosphere at a level that could prevent dangerous anthropogenic interference with the climate system. And it is already recognised that under most equity interpretations, developed countries will need to reduce their emissions significantly by 2020 (10 to 40% below 1990 levels) and to still lower levels by 2050 (45 to 90% below 1990 levels) for low to medium stabilization (IPCC 2007b:90; IISD 2008).

Hence, technology is expected to play a major role in the post-2012 climate regime, both as a means to attract action from major emitters, including large developing nations and those not currently party to the Kyoto Protocol and to ensure that countries have the necessary tools to put their economies on a clean development path. Technology will dictate what is possible in regard to emission reductions and technology development and transfer is identified as one of four pillars in the Bali Action Plan, the negotiating mandate agreed to at COP-13 (IISD 2008). More obviously, to reduce emissions while meeting growing energy needs will require the acceleration of technological advancement and a reduction of costs to encourage the wide uptake of zero- or low-emission technologies at the national and international level. Efforts will require the widespread use of all existing technologies and major efforts in energy efficiency, as well as continued efforts to lead to technology breakthroughs (e.g., nuclear fusion, new methods for hydrogen storage, new energy storage devices and new technologies for improving energy efficiency). Discovery and innovation in the energy sector is still capable of yielding unanticipated rewards with regards to GHG emission reductions (IISD 2008).

While energy technologies have been at the forefront, there have been efforts to cooperate on technologies to deal with non-energy emissions such as carbon dioxide from industrial processes; carbon dioxide from land use, land use change and forestry (LULUCF), methane and nitrous oxide emissions from agricultural practices and methane from solid waste landfills (IISD 2008). In the technological transfer and cooperation process, the developed country Parties are needed to support the development and enhancement of endogenous capacities and technologies of developing country Parties.

Consequently, the United States has stressed a technology approach in its climate change programmes and has been a driving force in the establishment of a number of multilateral technology agreements (i.e., APP, Carbon Sequestration Leadership Forum [CSLF], Generation IV [GENIV] on nuclear energy systems, Global Nuclear Energy Programme [GNEP], International Partnership for the Hydrogen Economy [IPHE] and Methane to Markets [M2M]). Other multilateral technology agreements include ITER, IEA Implementing Agreements, and the Renewable Energy and Energy Efficiency Partnership [REEEP]. Technology cooperation is also promoted through bilateral agreements to promote technology (e.g., EU-China Partnership on Climate Change, 2006 U.SA. - Russia framework for bilateral cooperation in the development of nuclear energy technology and deployment), and through bilateral Official Development Assistance (ODA) projects. Japan is the most significant donor of bilateral aid for energy, having provided 69& of all bilateral aid for energy during the period 1997–2005, followed by Germany at 12% and France at 3.4% (Tirpak & Adams 2007: 3) (IISD 2008).

Estimates suggest that billions of dollars will be required to assist countries in technology efforts. UNFCCC's (2007: 92) analysis estimates that additional investment and financial flows of US$200–210 billion are needed globally for mitigation to return CO_2 emissions to current levels by 2030. A technology approach to reducing fossil fuel emissions outlined in IEA's Energy Technology Perspectives Report (2006 in Stern 2006: 234) estimates a total net cost of US$100 billion over the next 45 years to bring energy-related emissions down near to current levels by 2050.

Stern (2006: 370) also suggested that in addition to a carbon price, deployment incentives for low-emission technologies should increase two to five times globally from current levels of approximately US$33 billion per year and that global government energy R&D budgets need to double to US$20 billion per year to ensure the development of a diverse portfolio of technologies. These figures, consistent with a 500 ppm CO_2-eq stabilization level, are considered to be modest when compared with predicted overall levels of investment in energy supply infrastructure up to 2030. In all cases, the absolute costs are high, although the Stern Review notes that mitigation costs will vary according to how and when emissions are cut. Without early action, the costs of mitigation will be greater (IISD 2008).

3.3 Funding Mechanism

Financing for climate change has generally flowed from four sources: 1) bilateral and multilateral development assistance, including the GEF; 2) the carbon market, including the CDM; 3) foreign direct investment (FDI); and 4) internally generated sources of funds, including government and private sector financing. Each of these sources is expected to continue to play a critical role in the future climate regime (IISD 2008).

During the period from 1997 to 2005, inflows of all FDI to developing countries totaled over US$2 trillion. Inflows of FDI to developing countries were US$267, 164 and 334 billion in 2000, 2002 and 2005 respectively, with FDI nearly three times higher than ODA in 2005. FDI tends to rise and fall with financial cycles and be risk adverse (UNCTAD 2006). It is also selective—FDI will only flow to those countries where relatively strong enabling conditions for investment exist. These include stable political environments, strong leg legal systems, macro-economic stability and available skilled labour and good institutions.

Since many of the poorest countries do not have these basic governance conditions, ODA remains an important source of funding for technology transfer for these countries (Ellis *et al.* 2007). It is difficult to obtain data on private sector investments for facilities and equipment in the energy and industrial sectors which may have an effect on GHG emissions. However, data are available from UNEP and New Energy Finance (2007) for sustainable energy investments, particularly renewables. These data indicate that investment in sustainable energy is rapidly increasing, with US$117 billion of new investment in 2007, which was 41% more than 2006. This amount beat a forecast of US$85 billion for 2007 by a wide margin. Additional insights from the UNEP and New Energy Finance (2007) about recent trends in financing for sustainable energy financing are listed below (IISD 2008).

Over the period 1997 to 2005, official development assistance (ODA) provided by donor countries for all purposes totalled approximately US$490 billion. This period in time saw a significant increase in financing, rising from the low level of US$60 billion in 1997 to US$106.8 billion in 2005, the highest level ever in real and nominal terms. The level of funding provided in

2005 was high due in part to exceptional circumstances; the Paris Club's debt relief effort for Nigeria and Iraq that accounted for nearly 20% of the total, and tsunami relief and other humanitarian needs (IISD 2008). However, it is difficult to determine the portion of this funding that was directly relevant to climate change mitigation and adaptation. While the OECD maintains a database of development assistance for disaster relief, education, health, energy, agriculture and other areas, most projects are initiated for non-climate reasons and often have multiple purposes (IISD 2008).

Bilateral and multilateral support for energy projects totalled over US$64 billion from 1997 to 2005, with multilateral institutions accounting for nearly 70% of the support. Support for energy projects has ranged from six to 10% of all development assistance during this period but was virtually stagnant at approximately US$6–7 billion per year in the same period (see Table 2). The World Bank Group was the largest source of multilateral funds, contributing nearly 39% of all funding, including bilateral funds. Support for the power sector, while down from earlier years, has dominated energy funding. Funds for the oil and gas category have been relatively constant, while support for energy efficiency measures and renewables has been variable, despite efforts since the early 1990s to expand both portfolios (IISD 2008). Collectively, the power, coal, oil and gas categories account for 75% of all funding. The GEF represented only 2% of all development assistance for energy. However, this may mask its importance to many small developing countries (Tirpak & Adams 2008; IISD 2008).

Bilateral development assistance for energy was at its highest in 1997, reaching nearly US$4 billion and at its lowest in 2000 (approximately US$1.3 billion). It has recovered in recent years, during which funding has averaged slightly more than US$2 billion annually. Japan provided over two-thirds of all bilateral aid for energy during the period 1997 to 2005, out-distancing by far the next most important donors (as seen in Table 2), namely Germany (12.0 %) and France (3.4 %) (IISD 2008)

Table 2: Multilateral and bilateral funding for energy during the period 1997–2005 (US$ millions)

SOURCE/YEAR	1997	1998	1999	2000	2001	2002	2003	2004	2005	TOTAL
BILATERAL ODA	3992	2522	1820	1294	1950	1950	2726	2296	2132	20682
WORLD BANK GROUP	3633	3833	2258	2683	2817	2817	2450	1828	2794	25113
EBRD	357	357	357	387	620	680	667	768	765	4958
GEF	136	113	83	113	97	97	120	134	124	1017
ADB	824	400	699	1042	663	927	654	707	677	6593
IADB	1131	1261	464	1172	1188	184	379	152	1056	6987
TOTAL	10073	8486	5681	6691	7335	6655	6996	5885	7548	65350

Source: *Tirpak and Adams, 2008; in IIED 2008*

There is some evidence that cash transfers can build assets or provide households with contingency finance for mitigating climate-related risks. However, the timing has to be right; both in terms of coinciding with the season when there is less food and also making sure the amount of transfer takes adequate account of purchasing power, which can vary over the course of a year (Tanner & Mitchell 2007; Davies *et al.* 2008).

Elements contained in the market mechanism similar to the Clean Development Mechanism (CDM), which allow developing countries to finance their efforts to simultaneously reduce GHG emissions and foster sustainable development, are not fully attained. There are significant differences in the proposals however, in terms of access by Annex I countries to REDD credits as well as calls for Annex I countries to finance developing country efforts without the reductions for target attainment purposes. Also missing from this are proposals from the major developing emitting countries that will inevitably seek to gain credit through REDD as they take on commitments over the longer term (IISD 2008).

In addition, the current levels of investment in adaptation measures are substantially less than those estimated to be needed in the relatively near future (IISD 2008). This situation includes the provision of financial resources from developed to developing countries to assist them in preparing for and responding to the unavoidable impacts of climate change. Since 2005, the international community has provided approximately US$11 million to developing countries for adaptation related activities through funds managed by the Global Environment Facility (GEF) (ibid). Additional bilateral contributions are estimated to have amounted to US$100 million between 2000 and 2003 (UNFCCC 2007a; IISD 2008).

Recognizing that some development assistance not ear-marked for adaptation can contribute to building adaptive capacity, these official contributions for adaptation still are significantly less than the billions of dollars potentially needed each year (IISD 2008). A long-standing difference among Parties is their perception of that basis upon which financing for adaptation should be provided. On the one hand, developing countries note the historic responsibility of developed countries for the problem of climate change and cite the need for compensation for the damages caused. On the other hand, developed countries have treated their financing of adaptation activities as development assistance unlinked to any explicit notion of direct responsibility. This dichotomy is likely to be a central point of disagreement in the post-2012 adaptation negotiations (see, for example, IISD 2008).

Linked to this debate is the desire to ensure greater predictability in the provision of future adaptation funding. At present, adaptation funding is provided by developed country governments through bilateral development assistance and contributions to the various climate change funds. The relatively limited funding made available through these channels can be tied to the voluntary basis upon which they are provided. Mandatory contributions, potentially based on historic levels of greenhouse gas emissions, could increase the amount and predictability of adaptation financing. However, tying mandatory contributions to historic or current emission levels implies that compensatory funding is being provided; contributions made on this basis are unlikely to be attractive to donor countries (IISD 2008).

Allocation of responsibility between developed and developing countries as previously noted, suggests that tens of billions of dollars will be needed each year to support developing countries' adaptation efforts. It is extremely unlikely that current funding mechanisms will be able to mobilize this level of financial resources (e.g., Bouwer *et al.* 2004; World Bank 2006a; Müller & Hepburn 2006; Müller 2007 in IISD 2008), especially as developed countries increasingly realize their own vulnerability to (and the associated costs of) climate change. Even if mandated contributions were introduced, it is unlikely that developed countries will agree to cover all and likely not even the majority of developing countries' adaptation costs. The "incremental cost" provision of the Convention (Article 4.3) itself implies that the initial burden for addressing the adverse effects of climate change rests with developing countries, even in situations where it is clear that measures are needed due to the impacts of climate change (Mace 2003 in IISD 2008).

Consequently, the significant portion of financing for adaptation might need to be derived from the budgets of developing country governments themselves (which will require integration of adaptation considerations into key domestic decision-making and funding channels). This scenario suggests that underlying future negotiations will be a debate regarding how much responsibility developing countries should bear for financing their own adaptation efforts (IISD 2008).

Capital is shifting to developing countries, which saw higher private investment in 2006. This reflects stronger FDI as well as private capital mobilization within emerging markets. China, India and Brazil are major producers of and markets for sustainable energy, with China leading in solar, India in wind and Brazil in biofuels. However, barriers to FDI remain, such as restrictions on foreign ownership in China, causing a prevalence of foreign and local joint ventures. Developing countries face the challenge of fast-growing energy demand combined with less mature capital markets—which skews investment towards conventional, mostly fossil-fuel generation (IISD 2008).

3.4 Challenges Associated with Compliance to Mitigation and Adaptation

While it is recognized that Annex I Parties have a responsibility to support developing countries with their adaptation efforts, the basis upon which this support is provided (assistance or compensation; voluntarily or compulsory) and the level of funding to be provided is a matter of considerable discussion. For example, as of September 2007, the United States had not contributed funding to either the Least Developed Countries Fund or the Special Climate Change Fund. In contrast, contributions to either or both of these funds had been received from 18 other developed countries, as listed in Annex 1 (GEF 2007b; IISD 2008).

Although actions and discussions to date under the UNFCCC have focused on technology transfer, this has proven to be a controversial topic. The controversy stems from the differing perceptions of what drives technology transfer; developing countries have called on developed countries to increase financial and technical support, focusing on the removal of intellectual property rights (IPR) and the creation of a new fund to buy patents. Developed countries have argued that the intellectual property does not belong to governments, but to the private sector and have pointed to the need to be creative to come up with some incentives for private companies that own the technologies (IISD 2008).

Multilateral development banks (MDBs) also contribute financially, with the World Bank (2006: 19) reporting that over the five year period to 2005, the World Bank Group (WBG), the African Development Bank (AfDB), the Asian Development Bank (ADB), the European Bank for Reconstruction and Development (EBRD), European Investment Bank (EIB) and the Inter-American Development Bank invested over US$17 billion in projects that directly or indirectly contribute to lowering carbon emissions in the developing countries and the EIB has invested close to US$30 billion in similar projects in the EU, European Free Trade Association and the EU accession countries. The World Bank notes that this is still a small portion of the overall resources required for clean energy.

Tirpak & Adams (2007) report that there is a considerable gap between current public funding and projected financing requirements for energy technology. While most of this gap may be filled by private capital, public funding, particularly grants, will be needed to reduce the risks associated with the introduction of new technologies and to encourage developing countries to implement more environmentally friendly, but more costly options (IISD 2008).

The increasing sustainable energy investments and growing international technology cooperation to deal with climate change are laudable, yet there is much work to do. In practical terms, very little transfer of hard technologies has taken place and technology cooperation agreements to date have not yielded substantial results (Ott 2007; Murphy *et al.* 2005; Republic of South Africa 2006 in IISD 2008)—certainly not enough to kick-start the deep reductions needed to stabilize carbon dioxide emissions at a safe level. Much remains to be done to promote the development and diffusion of climate-friendly technologies and an effective post-2012 agreement will need to include provisions to stimulate technology in developed and developing countries, far beyond what has taken place to date (IISD 2008).

3.5 The way forward

Governments have a key role to play in establishing the right environment for technology innovation and uptake and have a variety of options in the policy mix to instigate the development and deployment of zero- and low-emission technologies. There are various viewpoints on what is needed to initiate real action on climate change, although there is increasing consensus that both technology "push" and "pull" measures are needed to stimulate required changes (Fisher *et al.* 2006; Grubb 2005; Leach *et al.* 2005 in IISD 2008).

On the one hand, the technology push approach focuses on the development of low emissions technologies by governments or other public institutions playing a more direct role in funding technological change through such measures as R&D policies, support for demonstration projects, technology promotion programmes and research through effective public-private partnerships. Grubb (2005) noted that proponents of a technology push approach prefer investing in innovation in the short term and adopting emission reductions when technology improvements have lowered costs (IISD 2008). On the other hand, the technology pull approach focuses on technological change as a product that typically results from economic incentives and that such change will primarily come from the private sector in response to market signals or economic incentives. Examples of technology pull measures include corporate tax breaks for R&D expenditure, carbon taxes and emission trading schemes (IISD 2008).

A key consideration in encouraging large-scale technology transfer is the design of an international framework that can encourage appropriate price signals of similar magnitude in developed and developing countries. In market economies, relative prices must push in the correct direction or nothing of significance happens. Zero emission technologies can be developed with R&D and technology transfer can be fostered through mechanisms that affect a small percentage of the market— but the technologies will not be widely disseminated without appropriate price signals that account for the full social and environmental costs (negative externalities) associated with climate change (IISD 2008).

Adequate *financial support* will be needed to support technology development in and transfer to developing nations. Traditional patterns of financial support— ODA through bilateral and multilateral institutions— will remain important and be a key means for technology change in LDCs. Developed countries have to increase levels of ODA and to ensure that climate change and clean energy considerations are mainstreamed in programs and projects. The outcomes of the World Bank's clean investment framework could provide valuable lessons on effective aid programming. But efforts will need to go beyond ODA and be at levels that are greater than current ODA programming for climate change efforts. Increased funding commitments for technology from developed countries are likely to be part of a post-2012 agreement and this could include a technology fund (IISD 2008).

Effective adaptation must be locally driven and to a great extent, adaptation is a place-based activity, with planning and implementation of adaptation measures undertaken in response to local circumstances and capacity (e.g., Burton & van Aalst 2004; Downing *et al.* 2005). However, this is not to say that actors operating at other scales have no role to play; national governments, for example, have a key role in helping to create an enabling environment that facilitates adaptation. This can be achieved through capacity building, removal of barriers, knowledge sharing and financial support (Kartha *et al.* 2006).

Adaptation must be intimately connected with sustainable development. In all countries, sustainable development efforts can reduce vulnerability to climate change by tackling the fundamental factors that undermine adaptive capacity (IPCC 2007b). However, it is important to note that some response strategies do not translate into adaptation but rather maladaptation, for instance charcoal burning for trade (see Photo 11). Resilience to shocks and stresses, including climate change, can be built by strengthening institutions, promoting sound management of natural resources, improving health and education systems and fostering economic growth (AfDB *et al.* 2003 in IISD 2008). At the same time, it needs to be recognized that development activities that do not consider the probable impacts of climate change can be put at risk (e.g., introducing new tree crops in areas that will be climatically unsuitable over the long-term). In addition, development measures undertaken without the application of an "adaptation

Photo 11: A charcoal vendor searching for customers. Charcoal business seems to be another key source of income to supplement dwindling incomes of peasants. However, this process could well illustrate the vicious cycle of poverty.

lens" can undercut adaptive capacity (Adger *et al.* 2003; Orindi & Eriksen 2005 in IISD 2008). For example, development efforts can encourage people to live in increasingly hazardous areas or remove supports systems that are relied upon in times of stress.

Adaptation *requires an integrated approach.* Adaptation is a long-term, continual process—one that will involve all sectors of society. This characteristic, combined with its intimate connection to sustainable development, means that adaptation efforts cannot be effectively undertaken separately from ongoing development and governance processes. Integrating consideration of the impacts of climate change into every day decision and policy-making processes is more efficient and effective than addressing adaptation as a stand-alone issue. An integrated approach also is considered to be the most economically efficient way to address the consequences of climate change. By integrating consideration of the implications of climate change into financial and investment decisions, exposure to unacceptable risk can be reduced. Moreover, it is less likely that development projects undertaken today will inadvertently increase the vulnerability of communities over the long term (Huq & Burton, 2003; OECD 2005; World Bank 2006c; Klein *et al.* 2007; IISD 2008).

A fundamental challenge of climate change mitigation and adaptation is how to promote a technology revolution on a global scale that would enable, at acceptable costs, the steep reductions required to stabilize atmospheric greenhouse gas (GHG) concentrations at a safe level (IISD 2008). Attaining these global emission reductions require a significant transformation in the capital stock of energy producing and consuming businesses and households around the world. Almost every major societal function results in GHG emissions such as transportation, agriculture, space heating, manufacturing, forestry and social and economic systems.

The critical question is how best to engage countries in a long-term effort that will mobilize technologies to protect the climate while sustaining economic growth (IISD 2008) On the mitigation side, there is a need to design emissions-trading frameworks that support greater African participation than at present and that include land-use change. Mitigation undertaken elsewhere will have a major impact on Africa, both positive (e.g. new technologies) and negative (e.g. commodity price changes arising from biofuel policies) (IISD 2008).

CONCLUSION

The impact of climate change is already being felt in Africa and strategies for facilitating adaptation need to be developed. Parts of Africa stand to be particularly hard hit because of their geography, their agricultural dependence and because of difficulties that adaptation will face as outlined in this chapter. This remains unfair for the continent as climate change is not a problem of Africa's making. In this regard, a range of actions need to be pursued by African governments and by the international community. In the long term, world mitigation that is effective in stabilizing atmospheric levels of CO_2 will be valuable to Africa, as it will be valuable for the rest of the world. African adaptation to future climatic deterioration and opportunities for African participation in schemes for mitigation are key factors that need to be taken into cognizance. World mitigation efforts will also affect Africa indirectly via price effects and technical change. Production subsidies to OECD farmers can damage developing country interests. In addition, public financial support may be required in shaping new technologies so that they are applicable in Africa and developing regions more widely.

CHAPTER FOUR

LINKING NATURAL RESOURCES
AND CLIMATE CHANGE

4.0 Introduction

At the international level, the links between climate change and natural resources are well recognised. Within the UNFCCC[10], Article 2 emphasises that stabilization levels should be achieved within a time frame sufficient to allow ecosystems to adapt naturally, while in Article 4.1 all Parties commit themselves to protecting sinks and reservoirs. The Kyoto Protocol also emphasizes the importance of ensuring that carbon sequestration activities contribute to the objectives of the CBD[11]. For instance, the 8th CoP of the Ramsar Convention in 2002 called on parties to "take action to minimize the degradation as well as promote the restoration of those peat lands and other wetland types that are significant carbon stores" (Roe 2006). In addition, the 7th CoP of the CBD (2004) encouraged Parties to "take measures to manage ecosystems so as to maintain their resilience to extreme climate events and to help mitigate and adapt to climate change" (Roe 2006 p.32).

In general, natural resources are inextricably linked to climate changes. This is based on the grounds that climate changes affect natural resources such as land and biodiversity; and changes to natural ecosystems affect climate parameters (Mansourian *et al.* 2009; Reid *et al.* 2004; Reid 2004). Therefore, just as climate change affects natural resources, so changes in natural resources such as biodiversity can also affect the global climate. For instance, land use changes that lead to biodiversity losses can cause increased greenhouse gas emissions. Forests are a major store of carbon, and when forests are cut down or burnt, carbon dioxide is released into the atmosphere. For instance, continuing deforestation, mainly in tropical regions, is currently thought to be responsible for annual emissions of 1.1 to 1.7 billion tonnes of carbon per year or approximately one-fifth of human carbon dioxide emissions (Reid 2004).

This linkage acknowledges that effective biodiversity conservation and management can lead to higher levels of carbon sequestration and hence climate change mitigation. For example, forest management activities such as increasing rotation age, low intensity harvesting, reduced impact logging, leaving woody debris, harvesting which emulates natural disturbance regimes,

10 UNFCCC: United Nations Framework Convention on Climate Change
11 CBD: Convention on Biodiversity

Photo 12: Supporting developing countries to protect natural forests could well enhance efforts of reducing acceleration of climate change.

avoiding fragmentation, provision of buffer zones and natural fire regimes can simultaneously provide biodiversity and climate benefits. Forests (Photo 12) play a critical role in reduction of emissions. This is also true for certain agro-forestry, revegetation, grassland management and agricultural practices such as recycling and use of organic materials. Integrated watershed management can conserve watershed biodiversity in addition to increasing water retention and availability in times of drought; decreasing the chance of flash floods and maintaining vegetation as a carbon sink (Reid 2004).

Energy production is another key area where biodiversity conservation provides opportunities to help mitigate climate change. Currently, some 60% of anthropogenic global greenhouse gas emissions originate from the generation and use of energy. Use of renewable energy sources provides an opportunity to reduce emissions from burning fossil fuels (Reid 2004).

Poor people are therefore severely affected when the environment is degraded or their access to it restricted. This is attributed to the fact that poor people generally depend more on ecosystem services and products for their livelihoods than wealthy people. The means by which a poor family gains an income and meets its basic needs are often met by multiple livelihood activities. For example, exploiting common property resources such as fish, grazing land or forests can provide income, food, medicine, tools, fuel, fodder, construction materials and so on. As a result of this dependency, any impact that climate change has on natural systems threatens the livelihoods, food intake and health of poor people (Smith & Troni 2004; Reid 2004).

Climate change will mean that many semi-arid parts of the developing world will become even hotter and drier with even less predictable rainfall. Climate-induced changes to crop yields, ecosystem boundaries and species' ranges will dramatically affect many poor people's livelihoods. Those most vulnerable to climate change are the poorest groups in the poorest countries of the world. This is because they live in areas more prone to flooding, cyclones and droughts and also because they have little capacity to adapt to such shocks. They are often heavily dependent on climate sensitive sectors such as fisheries and agriculture and the countries they live in have limited financial, institutional and human capacity to anticipate and respond to the direct and indirect impacts of climate change (Walter & Simms 2002; Huq *et al.* 2003; Sperling 2003; Tyler & Fajbar 2009).

Conservation of biodiversity and maintenance of ecosystem integrity may be a key objective towards improving the adaptive capacity of such groups to cope with climate change. Functionally diverse systems may be better able to adapt to climate change and climate variability than functionally impoverished systems. A larger gene pool will facilitate the emergence of genotypes that are better adapted to changed climatic conditions. As biodiversity is lost, options for change are diminished and human society becomes more vulnerable (Reid 2004).

4.1 Climate change impacts and natural resources

Climate change is one of humankind's most pressing integrated economic, social and environmental issues (Parry *et al.* 2005). Climate change is projected to impact broadly across ecosystems, societies and economies, increasing pressure on all livelihoods and food supplies, including those in the fisheries and aquaculture sector (Cochrane *et al.* 2009). These threats of climate change to human society and natural ecosystems have been elevated to a top priority in the world (IPCC 2007; Cochrane *et al.* 2009). While all people and ecosystems are vulnerable to climate variability and change, the impacts are location specific. They depend on the nature of climate change and variability, the speed of the change, sensitivity of the area and the adaptive capacity of its people and ecosystems (FAO 2009).

Hence, poor rural people in most developing countries are already puzzled by shifting seasonal climate patterns which weaken the value of their long base of experience and local knowledge and expose them to new risks of food insecurity (Oxfam 2008; Tyler & Fajbar 2009). For instance, development challenges in coastal areas will exacerbate increases in frequency and intensity of tropical storms, higher sea levels and storm surges and coastal and beach erosion leading to loss of livelihoods, property and infrastructure. Both artisanal and large-scale fisheries and marine aquaculture will be heavily affected (Tyler & Fajbar 2009).

Coastal deltaic areas are especially vulnerable to climate change impacts due to the combination of more variable upstream flows, as well as coastal storms and sea level rise. Low gradients and intensive agriculture in these areas reduce drainage and encourage saline intrusion. Change in hydrological flow and flooding will affect crop production, fisheries and human health (Tyler & Fajbar 2009).

In terms of physical and biological impacts, climate change is modifying the distribution of natural resources such as marine and freshwater species. In a warmed world, ecosystem productivity is likely to be reduced in most tropical and subtropical oceans, seas and lakes and increased in high latitudes. Increased temperatures will also affect fish physiological processes; resulting in both positive and negative effects on fisheries and aquaculture systems, depending on the region and latitude (Cochrane *et al.* 2009).

Photo 13: Sustainable management of forests is important in managing climate change.

Today, climate change is one of the main emerging threats facing biodiversity. Up to a quarter of mammal species are at risk of global extinction because of climate change. Climate change is expected to cause species to migrate to areas with more favourable temperature and precipitation (IPCC 2002). There is a high probability that competing, sometimes invasive species, more adapted to a new climate will move in. Such movements could leave some protected areas with a different habitat and species assemblage than they were initially designed to protect (Mansourina et al. 2009).

The effects of climate change on river ecosystems are no longer just speculation (Ormerod 2009). Rivers have been sensitive to two indirect consequences of climate change. First, many are impaired already by other pressures with which climate interact. These include eutrophication, organic pollution, sediment release, acidification, abstraction, impoundment, urbanization, hydropower development, flood-risk management and invasion by exotic species (Ormerod & Durance 2009). Second, climate changes profoundly affect river conditions and processes indirectly by changing the human use of river catchments, riparian zones and floodplains (Ormerod 2009).

Water-related problems that already exist in the world are likely to worsen as a result of climate change. Intense rainfall events will increase the incidence of flooding in many areas. However, reduced runoff overall will exacerbate current water stress, reduce the quality and quantity of water available for domestic and industrial use and limit hydropower production. Access to water in the Nile basin countries is dependent on runoff from the Ethiopian highlands and the level of Lake Victoria, both of which are sensitive to variations in rainfall. While the impact of climate change on water scarcity may be relatively minor compared to socioeconomic changes such as increased demand, land cover change and economic growth strategies, it may have international consequences and become a source of conflict (Eriksen et al. 2008).

Sea level rise represents another threat to the region through saltwater intrusion and coastal erosion, although these effects will only be felt toward the end of the 21st century. Some of these climatic changes may have devastating effects where they add to existing stresses such as water scarcity and climatic variations such as decadal drying events. In addition, uncertainty regarding the direction and magnitude of changes in precipitation, river flows and lake levels in particular represents a challenge for adaptation to climate change (Eriksen et al. 2008).

Climate change and biodiversity loss are both major environmental concerns, yet the links between them often go unrecognised. Not only does the science of climate change and biodiversity share similar characteristics, but climate change both affects and is affected by biodiversity. Diversity confers far greater resilience on natural systems, thus reducing their vulnerability – and the vulnerability of the people that depend upon them – to climate change. Yet climate adaptation and mitigation strategies that are blind to biodiversity can undermine this natural and social resilience. Ignoring the links between biodiversity and climate risks exacerbates the problems associated with climate change and represents a missed opportunity for maximising co-benefits (Roe 2006).

Climate change is likely to have a number of impacts on biodiversity – from ecosystem to species level. The most obvious impact is the effect that flooding, sea level rise and temperature changes will have on ecosystem boundaries, allowing some ecosystems to expand into new areas while others diminish in size. As well as shifting ecosystem boundaries, these changes will also cause changes in natural habitat – an outcome which will have a knock-on effect on species survival. A growing body of research indicates that, as a result, climate change may lead to a sharp increase in extinction rates. In addition, literature shows that for many species, climate change poses a greater threat to their survival than the destruction of their natural habitat (Reid 2004).

The impact that floods, sea level rise and changes in climate are likely to have on natural habitats means that some protected areas may no longer be appropriate for the species they were designed to conserve (Reid 2004). Global warming is also causing shifts in the reproductive cycles and growing seasons of certain species. For example, higher temperatures have led to an increase in the number of eggs laid by the spruce budworm, already one of the most devastating pests in North America's boreal forests. However, the impacts of climate change on biodiversity will vary from region to region. The most rapid changes in climate are expected in the far north and south of the planet and in mountainous regions. These are also the regions where species often have no alternative habitats to which they can migrate in order to survive. Other vulnerable ecosystems and species include small populations or those restricted to small areas. Coral reefs have already shown devastating losses as a result of increased water temperatures (Reid 2004).

Increasing mean annual temperatures might initially promote greater forest productivity. As temperatures continue to rise though, evaporative demand is expected to increase while soil moisture decreases, leading to an increase in the frequency and intensity of drought. These changes are expected to impact each tree species differently. Some will be able to cope; others will not. Drought-stressed forests will be more susceptible to damage from insects and disease;

climate change may lead to an increase in the frequency and intensity of insect, disease and fire events (FAO 2009). Despite forests contribution in the enhancement of resilience capacity to both human and natural systems, forest ecosystems may not be able to adapt to the rate of temperature change or the intensity of weather events and other effects such as fires or floods (IIED[12] 2009).

Finally, land management will increasingly be affected by climate change and its many socio-economic consequences. These include global food security, fuel security, water scarcity, population displacement and management for carbon sequestration. These consequences will likely drive agricultural change, intensification, forestry practice, water resource development and other land-use patterns over extensive areas. Not only will the direct demands on land use and management change per se in areas already under production, but also the geographical distribution of land uses will change as water scarcity increasingly limits options. In particular, arid and highly populated areas of the world that are unable to increase water supply or agricultural production will increase their demands for food exports from other regions (Ormerod 2009).

4.2 The role of forests in climate change mitigation

Forests cover 30% of the total land surface of the world (FAO 2007). Forests in the ten most forest rich countries account for two-thirds of total forest area, while 57 countries have less than 10% of their land area in forests (ibid). However, many existing forests and most newly established stands experience climatic conditions that deviate from conditions today (FAO 2007). In the context of climate change, mitigation refers to a human intervention to reduce the "sources" of greenhouse gases or enhance the "sinks" to remove carbon dioxide from the atmosphere (Chandler et al. 2002). These efforts of reducing carbon emission are mainly focused on enhancing absorption channel by conserving and restoring forest resources.

Despite the extreme climate changes impacts such as drought and increased floods that threaten forest resources, the roles of forests on human and natural systems are still enormously diverse; they range from a pivotal role in traditional agro-forestry systems by providing shelter, shade and protection against the ravages of wind, salt spray and sun. The diversity of functions in forest resources is particularly evident through mangrove forests and other coastal trees which play multiple roles in protecting coastlines, buffering wind and wave action and contributing to food webs (FAO 2009). Mangrove forests also play a vital but often undervalued cultural and ecological role in many coastal communities. In addition to mitigating coastal erosion, salt spray

12 IIED: International Institute for Environment and Development

incursion and coral siltation, mangrove forests provide protection from storm surges and tsunamis and provide important habitats for a wide variety of bird, crab and shellfish species. Likewise, mangroves form important habitats and nurseries for numerous pelagic and coastal fish species, many of which form a vital source of protein for island communities and coastal dwellers (FAO 2009).

In relation to climate change mitigation, forests can play a role in adaptation by helping human societies to stabilise resilience capacity in adapting to climate change impacts. It is estimated that adaptive management of forests contributes to sustaining the livelihood of over two billion people worldwide (FAO 2007). However, many existing forests and most newly established stands experience climatic conditions that deviate from conditions today (ibid). Forests may also serve as a source of resilience by absorbing harmful carbon dioxide emissions, providing resources to local populations and through forest-landscape design to protect communities from increasingly erratic weather. Hence, it is acknowledged that forests have substantial contributions to national and global mitigation portfolios designed to reduce the rate of carbon dioxide (CO_2) increases in the global atmosphere (Larsson *et al.* 2007).

For example, the objective of UNFCCC (Article 2) is to stabilize greenhouse gas (GHG) concentration in the atmosphere at a level that would prevent dangerous anthropogenic interference with the climate system within a time frame sufficient enough to allow ecosystems to adapt naturally to climate change. This may ensure that food production is not threatened and also ensure that economic development proceeds in a sustainable manner (Singh 2008).

This calls for the reduction and stabilization of atmospheric carbon dioxide concentrations at acceptable levels by 2100. Although there is no legal requirement that carbon dioxide be stabilized at this acceptable level; bringing the concentration as close as possible to the pre-industrial level is what the society must aim at in the long run (Pandey 2002; Singh 2008). Since carbon (C) emissions from deforestation and degradation account for about 20% of global anthropogenic emissions; deforestation has been accounted as the single largest source of land-use change emissions, resulting in extreme carbon emissions. Estimated net annual decline in the forest area globally in the 1990s was 9.4Million hectares (Mha), representing the difference between the annual deforestation of 14.6Mha and the annual afforestation of 5.2Mha (FAO 2001; Singh 2008).

Hence, the Stern Review (2006) reinforces the finding that forest conservation, afforestation, reforestation and sustainable forest management can provide up to 25% of the emission reductions needed to effectively combat climate change. The Review concludes that curbing deforestation has the potential to offer significant emission reductions fairly quickly in a highly cost-effective manner (Singh 2008). These payments can occur either for carbon sequestration (deriving from the net absorption of carbon dioxide in planted trees) or by protecting carbon stocks – which would otherwise be emitted – in natural forests. If the providers of ecosystem services can be fairly rewarded there is a good chance of reducing tropical deforestation and mitigating greenhouse gas production. These ideas are being implemented in pioneering efforts around the world. The challenge ahead is to replicate, scale up and sustain these pioneering efforts. This requires major advances in the scientific understanding of natural capital, as well as in the design and implementation of finance mechanisms and supporting policies and institutions (Bond *et al.* 2009).

Accordingly, international cooperation to assist developing countries in preventing deforestation through carbon trading is now regarded as one essential vehicle for mitigating the impacts of global warming (Stern 2006; Hall 2008). While no panacea, it is increasingly seen as one viable policy option if appropriately conceived and implemented. Yet neither should Payment for Ecosystem Services (PES), such as the reduced emission from deforestation and degradation (REDD) be viewed as a plain success. Many problems must be overcome if its potential is to be realised (Hall 2008).

Decisions taken at the 2007 Conference of the Parties to the UNFCCC in Bali (COP 13) reopened the possibility for REDD to become part of a post-2012 global climate regime. Consequently, a number of developed-country governments and international development agencies have forged high-level partnerships and allocated substantial new funds to help prepare countries for participation in a REDD regime, including support for capacity-building, pilot demonstration activities and other policies and measures to achieve reduced forest emissions (Bond *et al.* 2009).

While debate grows over the international architecture of a REDD mechanism and negotiations continue in various UNFCCC fora, more attention has been focused on how performance-based payments and other approaches to REDD (Bond *et al.* 2009) would operate at national and local levels; the priorities for up-front investment to strengthen country capacity to implement REDD and how REDD mechanisms can be designed to maximise co-benefits for forest-dependent communities and biodiversity conservation (Bond *et al.* 2009).

Therefore, in developing appropriate management strategies involving forests, managers are increasingly expected to consider a wide range of issues and indicators, including the impacts of their actions on the greenhouse gas balance (Larsson et al. 2007). For instance, appropriately designed CBFM[13] policy can provide means to sustain and strengthen community livelihoods and at the same time avoid deforestation, restore forest cover and density, provide carbon mitigation and create rural assets. Probably channelling carbon investment funds into CBFM projects can make both development and conservation economically viable and attractive for the local communities to maintain biodiversity and integrity of nature (Singh 2008).

Forest conservation for carbon sequestration purposes can be either direct or indirect. Direct interventions essentially require the "locking up" of threatened land resources into untouchable preserves. Indirect interventions comprise a far wider range of possibilities, including increasing agricultural productivity (thus lowering the need for cyclical slash and burn cropping), the development of agro-forestry to meet fuel wood needs, the opening of markets for indigenous forest products and the promotion of wood waste and paper recycling (Singh 2008).

Also, existing literature has noted that forests sequester and store large amounts of carbon in biomass and soils. Management practices, including afforestation, reforestation, and harvest can substantially influence carbon sequestration potential of the land. Therefore, forest mitigation strategies may involve eliminating forest land conversions (especially in the case of tropical deforestation), postponing harvests, reducing burning or increasing carbon uptake through intensified forest management and conversion of agricultural land to forestry (Alig 2003). Appropriate management approaches are currently in place and technologies are moderately available (Siry *et al.* 2009).

Thus, forest management techniques can enhance or reduce the effects of climate change. Consequently, it is widely acknowledged that CBFM micro-planning exercise at the decentralized and site-specific level calls for involving the indigenous communities and their prescriptions for managing and restoring forests. It can simultaneously be used to mitigate and overcome proximate threats of fragmentation and degradation and at the same time manage forests in such a way that the resilience and resistance of forests to climate change is enhanced. By protecting and restoring biodiversity, providing connectivity, mimicking nature in plantations and controlling man-made fires, CBFM is an effective way of managing forests during climate change (Singh 2008). In addition, there are other mechanisms that can be useful in minimising carbon emission through proper management of forests. The suppression of forest fires is one option to reduce unnecessary carbon emissions. Along with the crucial need to address the policy causes, a combination of ground-based practices of

---

13 CBFM: Community Based Forest Management

fire prevention and control has great potential for reducing the frequency and extent of forest fires (Stuart & Costa 1998).

Finally, forestry can also be used to prevent carbon released by fossil fuels elsewhere. This can be achieved through fuel or material substitution. Fuel wood can be used to replace fossil fuels and wood-based materials could be used to replace materials that require high levels of energy and/or fossil fuels for their production (e.g. steel, cement, plastics) (ibid).

4.3 Challenges in the use of forests for climate change mitigation

Although ecosystems play a significant role in regulating global climate; changes in biodiversity can affect this regulatory system. Forests, for example, are both a source and sink of carbon: carbon dioxide is fixed through photosynthesis but released into the atmosphere if forests are felled or burned (Roe 2006). While it is apparent that forests, and sustainably managed forests in particular (see Photo 13 for an example of a dense forest), can play an important role in climate change mitigation through increased uptake and storage of carbon dioxide, several important hurdles will have to be cleared before forests can fulfil their potential. These challenges range from developing effective forest carbon sequestration rules to compliance requirements and market considerations. They arise from the nature of carbon sequestration projects and management characteristics of the forestry profession (Siry *et al.* 2009).

Another important issue is that of permanence. As already stated, forest carbon sequestration is temporary. While carbon may be stored for decades in forests and wood products, eventually, it will be released. While it is possible to develop large forest projects which in due time will allow tree mortality and harvest to be offset by regeneration and growth, resulting in a steady state non-declining carbon pool, this may not be a very efficient approach to land use management (Siry *et al.* 2009).

Although the Kyoto Protocol clearly recognizes the role that forests and forest management play in reducing carbon dioxide emissions; it also places several restrictions on how this can be achieved. These restrictions are related to the principles of baseline, permanence, additionality and leakage. Their reach is much farther than the Kyoto Protocol itself as many of carbon emissions reduction schemes assume similar approaches. For instance, the Kyoto Protocol states that only forests established after 1990 are eligible for carbon offsets. And several other trading schemes assume the same base date (baseline) (Siry *et al.* 2009).

This may represent a problem for regions where a large amount of afforestation took place prior to 1990, effectively making these projects ineligible for carbon credits. The Protocol also requires that forest carbon capture projects demonstrate additionality. A carbon emission reduction is additional only when it was developed exclusively for the purpose of climate change mitigation. Projects implemented under business as usual or required by other laws and regulations are not considered additional. Determining the usual management practices in the real world often is quite difficult (Siry *et al.* 2009).

Further, the Protocol assumes and requires that carbon emission reductions are permanent. This reflects that carbon dioxide is removed from the atmosphere forever. Forest carbon sequestration, however, by its very nature is temporary. Although trees can store carbon for several decades, eventually the trees will die and the carbon will be released. What is the value of the temporary storage of carbon? There is probably some value in temporary storage as it will, in the least, provide more time for the development of alternative, permanent carbon emissions reduction technologies (Siry *et al.* 2009).

Another question is how carbon should be valued and how harvesting and wood product manufacturing should be treated. Carbon can be stored in wood products for many years, but many carbon schemes do not consider tree harvests nor do they allow credit for carbon stored in forest products. The answer to this question is critical for managed forests and the role they may assume in climate change mitigation. All these requirements may seem reasonable at first, but in practice they effectively remove managed forests from climate mitigation efforts. While managed forests will continue to sequester carbon and provide certain storage benefits, their true potential to increase carbon sequestration above the current, natural (without extra management effort) levels may never be realized (Siry *et al.* 2009).

Furthermore, assessing how much deforestation is being "avoided" can be a complex and controversial endeavour, which relates to social and economic aspects of a particular region. Often, government policies induce pressure on standing forests by specifically encouraging forest utilisation. Some countries view conservation as patrimonial and an affront against a nation's sovereignty. As such, there has been some negative bias among potential funders against the idea of resource "lock-ups", although several programmes have combined conservation with sustainable utilisation and other economic activities (Stuart & Costa 1998).

4.4 Implications of climate change adaptation for natural resources integrity

Adaptation to climate change is concerned "with adjustment in natural or human systems in response to actual or expected climatic stimuli or their effects, which moderates harm or exploits beneficial opportunities" (IPCC 2007). Given the current extreme impacts of climate change, adaptation to environmental variability has been undertaken (to varying degrees of success) by people for millennia. Farmers' adaptation to their environment, livelihood diversification and coping strategies to deal with the overall variability of their social and natural environment are well documented (Grist 2008). Throughout the world, the need to adapt to the effects of climate change has become increasingly evident. Several research programmes have raised awareness of the challenges that the climate changes process poses, particularly for poor developing countries (Tyler & Fajbar 2009). This awareness has resulted in increased commitment to support for adaptation at the global and international levels (ibid).

In addition, available studies have identified the most vulnerable countries and regions; adaptive capacity has been assessed and improved; and national action programmes have been initiated in many countries. Despite these achievements, climate change adaptation has not been adequately integrated into planning and policy in key natural resource management sectors (ibid). Currently, a number of innovative responses are already taking place at the local level to tackle the dual agendas of poverty alleviation and resource sustainability. These responses provide a base on which to enhance resilience to future impacts of climate change and are in many cases already responding to climate variability. Positive local efforts include strengthening production systems, building economic assets, improving access to markets and information, diversifying to less climate-sensitive livelihoods, reducing disaster risks through local planning and preparation and building foundations for all of these initiatives through more effective institutions of local governance and resource management (ibid).

Technical approaches to strengthening production systems are widely recognized as necessary but not sufficient to enhance resource sustainability and equity (Sayer & Campbell 2004). Attention to more equitable tenure arrangements and access to productive resources is essential for sustainability and for poverty reduction. Ultimately, community-based resource management institutions (water, forests, rangelands) strive for more participatory and locally responsive planning and aim to be inclusive of voices of poor and more marginal groups. In addition, innovative approaches to co-management have supported more equitable tenure arrangements and addressed conflicts between different pastoral and sedentary groups (Tyler & Fajbar 2009).

Local adaptation strategies have adopted integrated planning at the local level, across sectors and with attention to livelihoods as well as resource sustainability. Integrated management strategies that employ ecosystem and landscape approaches has been one step towards these frameworks and practices that recognize the interdependencies of land and water resources and the need to consider these resources more holistically (ibid).

Also, the concept of sustainable adaptation has emerged from an awareness that adaptation can have unintended negative effects both on peoples and on the environment and that there is a need to qualify exactly what types of adaptation are desirable (Eriksen 2009). The environmental sustainability aspect emphasises that adaptation needs to not endanger the environmental or economic integrity, neither for other groups at present or for future generations (see Photo 14). Therefore, mitigation of greenhouse gases becomes an important part of sustainable adaptation criteria. In high emission societies in particular, adaptation to climate change needs to take place in a way that does not increase emissions and hence aggravate the vulnerability of others. Increased use of air conditioning as a response to rising temperatures may not be a sustainable form of adaptation, for example. From focusing on local level development type measures, sustainable adaptation has come to have global significance (Eriksen 2009).

Photo 14: Proper management of forests is key to managing extreme conditions that may aggravate the vulnerability of living organisms.

Forests conservation depict a significant link between biodiversity and climate change, as they represent a defence against atmospheric carbon dioxide build up as well as a repository of the genetic heritage of the world's flora and fauna. From this perspective, reforestation of degraded forests is a remedy to solve climate change problems. Several studies have demonstrated increased tolerance to environmental extremes and greater temporal stability and recovery potential as species richness increases. In turn, species richness enhances stability through redundancy provided by multispecies membership in critical functional groups (Singh P.P 2008).

A functional group with more diverse membership can maintain its role in the ecosystem despite fluctuations in its member species. With interest in availability of a wide diversity of resources within their resource catchments, indigenous people also contribute to the restoration of biodiversity in depleted landscapes. Where a stake is created for them as is done in CBFM, their detailed knowledge of succession and habitat preferences of different species greatly contributes to such a process (Singh 2008). If biodiversity is maintained, long-term viability increases in case of global climate change because out of a multitude of native species, at least some individuals may respond better than the others (ibid).

CONCLUSIONS

Generally, forest carbon sequestration is one of the key approaches to reducing atmospheric carbon concentrations. Therefore, forestry can help mitigate climate change through afforestation, reforestation, avoided deforestation, silvi-cultural change, biofuels and carbon storage in wood products. This chapter has highlighted the role that forests can have in mitigating climate change. With climate change riding high on the political and economic agenda, more and more attention is being paid to different mechanisms for offsetting, reducing and preventing carbon releases into the atmosphere. The growing market for carbon offers great opportunities for linking greenhouse gas mitigation with conservation of forests and biodiversity and the generation of local livelihoods. Therefore, the new generation of carbon funds must address the need for a sustained reduction in carbon emissions, while also building good governance and strengthening the resilience and adaptive capacity of ecosystems and local communities in the face of increased vulnerability to climate change. To tackle climate change effectively, we need to "join the dots" between biodiversity loss, local livelihoods and land use changes such as deforestation. There is a strong need for credible standards that link curbing emissions with forest conservation to ensure they provide robust carbon benefits while incorporating biodiversity conservation and benefits to local communities. Conservation-based strategies that address carbon emissions, which include afforestation, reforestation and curbing deforestation must be made robust.

In this regard, performance payments, whether market- or fund-based, will be an important element of national and sub-national carbon mitigation strategies, such as the REDD mechanisms. However, if certain up-front conditions are not met, it is unlikely that Payment for Ecosystem Services (PES) will be an effective instrument for REDD. These up-front conditions include economic, institutional, informational and cultural conditions (Wunder 2008b). In these cases, investments in improved governance structures or other enabling policies and measures are more effective (Bond *et al.* 2009).

Hence, a PES approach to REDD requires effective and equitable governance frameworks and systems, such as clarity of land rights and functioning monitoring to enable the enforcement of conditionality and quid pro quo payments. However, in many areas where deforestation and degradation are at their highest, governance is weak and is an underlying cause of deforestation and forest degradation. Importantly, governance can vary considerably across a single country (e.g., Brazil) (Bond *et al.* 2009). This suggests that REDD interventions will need to be paced and sequenced in accordance with capacity-building achievements and thus the level of urgency for the forest governance.

PART
2

CASE STUDIES ON VULNERABILITY
AND ADAPTATION TO CLIMATE CHANGE
AND VARIABILITY

CHAPTER FIVE

LOCAL COMMUNITY BASED ADAPTATION TO EXTREME EVENTS
Experiences from East Africa

5.0 Introduction

Changes in the mean temperature, rainfall patterns and rainfall variability in Africa are extending dry seasons and increase the severity of periodic droughts. For instance, this is common in the interior part of Tanzania, which experiences higher temperatures and reduced rainfall. The northeast, southeast and the Lake Victoria basin is less exposed to droughts but these areas are likely to experience more frequent and severe flooding. The predicted sea level rise of 0.10-0.90 metres will aggravate flooding in the coastal areas such as Dar es Salaam; Mtwara and Lindi regions (see Clark *et al.* 2003; Hulme *et al.* 2001; IPC 2001; Mwandosya *et al.* 1998:1-24; Paavola 2003).

Furthermore, the impacts in East Africa are accelerated by the interrelationships between climate change, water and poverty that are intimately intertwined in the developing world (for example, Tanzania, Kenya and Uganda) where climate change is experienced primarily as increased variability in rainfall and availability of water. Other studies have also confirmed that climate change, especially changes in the availability of water – has the potential to aggravate adverse health outcomes, malnutrition and poverty as well as vulnerability to all forms of environmental stress (Paavola 2003).

More frequent floods will likely destroy infrastructure, buildings and belongings in the floodplains, which in urbanised areas are often populated by poor households. For example two-thirds of Dar es Salaam's population (over two million people) are considered to live in flood-prone areas (UNEP 2002: 241; Paavola 2004). Flooded pit latrines pollute surface waters and wells with human wastes and increase the incidence of water-borne diseases such as diarrhoea, typhoid and cholera. Moreover, warming, flooding and increased rainfall increase the spread and incidence of insect-borne diseases such as malaria (ibid).

On the other hand, droughts will impact all settlements, requiring more time for water collection and resulting in reduced water use. This impairs hygiene and contributes to the spreading and increased incidence of all contagious diseases (Cairncross 2003; IPCC 2001; Johnstone *et al.* 2002; McMichael *et al.* 1996; Patz *et al.* 2002; Paavola 2004).

5.1 Impacts of Drought

Droughts is one of the most significant (and frequent) environmental stressors for agriculturalists and pastoralists alike in East Africa (Paavola 2003; IDRC/ CCAA 2009). Although drought is a slow onset hazard, it is impacting on more people in Africa than most hazards. Worse enough, climate change is projected to increase the risk of drought over many parts of Africa in the 21st century (IDRC/CCAA 2009). Recent research also suggests that warm sea surface temperatures are thought to be responsible for the recent droughts in equatorial and subtropical Eastern Africa during the 1980s to the 2000s (Nelson & Stathers 2009). Drought diminished water supplies reduce crop productivity and have resulted in widespread famine in East Africa (IDRC/CCAA 2009).

Although drought is often regarded as natural event, it is postulated that impacts are often aggravated by poor policies or alternatively, conflicts over limited water, food and grazing resources (IDRC/CCAA 2009). This may also be attributed to the climatic condition in East Africa that is largely semiarid and characterized by high seasonal rainfall variability (Washington & Preston 2006). Hence, droughts occur frequently and are severe, impacting negatively on the sub-region's economic performance since more than 50% of the Gross Domestic Product (GDP) is derived from rain-fed agriculture (Jury 2002). The rain-dependent economic structures make the sub-region particularly vulnerable to the recurrent droughts, which lead to crop failure and a decrease in production (Vogel 1994).

Consequently, in the adjoining drier areas, food crop production is either marginal or not viable due to an insufficient length of moisture growing period, high rainfall variability and frequent occurrence of severe drought (Challinor *et al.* 2007; Sidahmed 1996; Mortimore 1998; SEI[14] 2008). On the other hand, the impacts of climate change on livestock are also felt from an increased severity and frequency of drought (see Photo 15). Deterioration of pastures during droughts and periods of over-grazing have resulted in poor health and death of livestock, which impacts food and livelihood security of those who own livestock (Challinor *et al.* 2007).

These impacts have threatened pastoral societies in Kenya, Tanzania and Uganda. In times of water scarcity, when livestock are forced to use the same water resources as humans, diseases are transferred between humans and animals and vice versa (SEI 2008). It is wise to acknowledge that the socio-economic impacts of droughts may arise from the interaction between natural conditions and human factors, such as changes in land use and land cover

14 SEI: Stockholm Environmental Institute

Photo 15: Prolonged droughts have adverse effects on sustainable livelihoods among pastoralists.

and water demand and use. Excessive water withdrawals can also exacerbate the impact of drought (Kundzewicz et al. 2007). Furthermore, economic development, increased urbanization and rapid population growth are likely to reduce per capita water availability throughout Africa and climate change is expected to exacerbate this situation, particularly in the seasonally dry areas (Cooper 2004; IPCC 2001).

5.1.1 Drought Experience in Kenya

In northwest Kenya, recurring drought periods have led to increased competition for grazing resources, livestock losses and conflict. For example, the recurrent droughts render almost impossible the problem of uplifting the standard of living of the population. For instance, during the last drought of the years 2005/6, the Maasai communities in the Kajiando district in Kenya lost 95% of their livestock. This rendered the whole population food insecure (IDRC/CCAA 2009; SEI 2008).

5.1.2 Drought Experience in Tanzania

A study by Ngana (1983) on drought and famine in Dodoma District indicated that the presence of dry spells in critical periods for most crops contributed considerably to crop failure and famine. Given the over-dependence on rain-fed agriculture by the majority of people living in rural areas, climate change and variability has been one of the major limiting factors in agriculture production thus resulting in food insecurity and low-income generation. For that reason, droughts have been reported to cause failure and damage to crop and livestock leading to chronic food shortages (Kangalawe & Liwenga 2005; Rosenzweig *et al.* 2002; Gwambene 2007; Mary and Majule 2009).

In Tanzania's semi-arid lands, availability of water for dry-land and irrigated cropping and other uses including domestic use and for livestock is currently threatened by excessive drought. The areas likely to be most hard-hit by a net decrease in rainfall are Tabora, Dodoma, Rukwa and southern Mbeya regions (CARE 2006). Rainfall is highly variable and unpredictable across the semi-areas of Tanzania contributing to risk and uncertainties of agricultural production activities and long-term temporal trends are weak or non-existent. Rainfall shortages regularly have led to serious droughts and eventually impacts on livelihoods, competition and conflict over natural resources. Also, high temperatures and less rainfall during already dry months in the Tanzanian river catchments affect the annual flow of rivers, such as the Pangani by reductions of 6-9% and the River Ruvu by 10% (VPO-URT 2003). For instance, the Pangani Basin being most agriculturally productive and an important hydropower production region; its decline due to climate change threatens the productivity and sustainability of this region's resources, which host an estimated 3.7 million people (VPO-URT 2003).

For instance, the gradual, yet dramatic disappearance of glaciers on Mount Kilimanjaro is a result of global warming (IPCC 2007). It is estimated that 82% of the icecap that crowned the mountain when it was first thoroughly surveyed in 1912 is now gone. According to recent projections, if recession continues at the present rate, the majority of the glaciers on Mount Kilimanjaro could vanish in 15 years (IDRC/CCAA 2009). The snow and glaciers of Mount Kilimanjaro act as a water tower and several rivers are drying out in the warm season due to the loss of this frozen reservoir. Other glaciers in East Africa which are under the same threat include Ruwenzori in Uganda and Mount Kenya (IDRC/CCAA 2009). This trend of ice melt due to temperature changes has resulted into declining of moisture needed for pastoral and agricultural activities and the availability of water for human consumption (Orindi & Murray 2005). Currently, two-thirds of rural Africans and a quarter of urban dwellers in Africa lack access to clean, safe drinking water (Simms 2005). In Tanzania, for example, two of three rivers have reduced flow due to declining regional rainfall, which has had ecological and economic impacts such as water shortages, lowered agricultural production, increased fungal and insect infestations, decreased biodiversity and variable hydropower production (IPCC 2001; IDRC/CCAA 2009; Orindi & Murray 2005). The Pangani Basin is also fed by the glaciers of Kilimanjaro, which have been melting alarmingly fast and are estimated to disappear completely by 2015 - 2020 (IDRC/CCAA 2009; Mwandosya *et al.* 1998:34-50).

Farming practices such as rainwater harvesting have been commonly used to cope with soil-moisture constraints (Hatibu *et al.*, 2000). It is, therefore, essential that we acquire greater knowledge of the climate processes that initiate and sustain these droughts and hence, the development of robust forecasting schemes (ibid).

5.2 Impacts of Floods

Globally, the number of great inland flood catastrophes during the last 10 years (between 1996 and 2005) is twice as large, per decade, as between 1950 and 1980, while economic losses have increased by a factor of five (Kron & Bertz 2007). The dominant drivers of the upward trend in flood damage are socioeconomic factors, such as increased population and wealth in vulnerable areas and land-use change. Floods have been the most reported natural disaster events in Africa, Asia and Europe and have affected more people across the globe (140 million/yr on average) than all other natural disasters (WDR 2003, 2004).

There is now a better understanding of flooding as a natural hazard and how climate change and other factors are likely to influence coastal flooding in the future (Hunt 2002; Boko *et al.* 2007). However, the prediction of precise locations for increased flood risk resulting from climate change is difficult, as flood risk dynamics have multiple social, technical and environmental drivers (Few *et al.* 2004b). The population exposed to flooding by storm surges will increase over the 21st century. Asia dominates the global exposure with its large coastal population: Bangladesh, China, Japan, Vietnam and Thailand having serious coastal flooding problems (Mimura 2001). Africa is also likely to see a substantially increased exposure, with East Africa (e.g., some parts of Tanzania, Kenya and Mozambique) having particular problems due to the combination of tropical storm landfalls and large projected population growth in addition to sea-level rise (Nicholls 2006, 2007).

Photo 16 illustrates flooding in a city.

The most affected and vulnerable groups are the vulnerable poor, especially women, children, elders, sick people, communities living in flood plains, communities in areas with poor infrastructure, areas with less social networks and communities living in flood plains (Majule 2009; Confalonieri *et al.* 2007). Likewise, the impacts of flooding will be felt most strongly in environmentally degraded areas and where basic public infrastructure, including sanitation and hygiene, is lacking. This will raise the number of people exposed to water-borne diseases (e.g., cholera) and thus lower their capacity to effectively use food (Easterling 2007; Boko *et al.* 2007; Freeman & Warner 2001; Mirza 2003; Niasse *et al.* 2004; Reason & Keibel 2004).

Generally, flood events have grown and are projected to occur much more in future to surpass population or economic growth, suggesting a negative impact on local and national development (Mills 2005). However, literature has not indicated clear trends for future flood occurrences (Kundzewicz *et al.* 2005; Schiermeier 2006). On the other hand, the observed increase in precipitation intensity and other observed climate changes, e.g., an increase in westerly weather patterns during winter over Europe, leading to very rainy low-pressure systems that often trigger floods (Kron & Bertz, 2007), indicate that East Africa might face greater impact of floods (Kundzewicz *et al.* 2007).

5.2.1 Experiences of Floods in Tanzania

It is estimated that about 38% of the past disasters in Tanzania have been caused by floods (Hatibu 2007). The most recent flooding occurred during the 1997/98 El-Nino rains. As a result of these floods, it was estimated that nearly US$ 200 million worth of damage to the infrastructure occurred (Hatibu 2007; Eriksen 2009). Accordingly, the northern Tanzanian bimodal rainfall regions (Arusha, Tanga and Kilimanjaro) were severely affected by the OND 1997 flooding and also in OND 2006 (Kijazi 2009).

In these regions, rainfall was unevenly distributed with recurring intense wet spells associated with widespread floods. The result was devastating loss of life and property, including the destruction of homes and properties, the loss of crops, damaged roads and bridges and the outbreak of diseases [such as rift valley fever (RVF)]. According to the International Federation of Red Cross, about 2900 households (13,522 people) were affected by the floods; and 200,441 cattle, 81 945 goats and 37,773 sheep had RVF symptoms. In addition, 174 people were admitted to hospitals with suspected RVF symptoms (34 died of the disease) (Kijazi 2009).

Although, ironically, it is assumed that the most devastating floods are caused by heavy rains which occur after long periods of drought, there have been unpredictable floods in semi-arid regions of Tanzania (Hatibu 2007). For example, despite the overall low amount of rain in the semi-arid areas, sometimes it falls in intense storms that often lead to flooding. This trend has attracted several scholars to consider flooding in Tanzania as a seasonal phenomenon occurring in localized areas (Hatibu 2007).

5.2.2 Experiences of floods in Uganda

In the last few decades, Uganda has experienced an increase in the frequency and intensity of extreme weather events with serious socioeconomic consequences. The most extreme event experienced being the El Nino of 1997/98. This El Nino is reported to have inflicted perhaps the most severe losses. For instance, it swept bridges - many towns were cut off from commercial centres causing heavy losses in goods and services- crops were destroyed and there was an outbreak of water borne diseases such as cholera and other flood-related diseases (Ben 2005). Following the El-Nino rains of 1997-98, an estimated 525 people died and over 11,000 were hospitalised and treated for cholera triggered by the El Nino induced floods and land slides. Again, an estimated 1,000 people were reported to have died in flood related accidents and about 150,000 people were displaced from their homes. Damage to trunk and rural roads infrastructure was estimated at $US 400 million. In Kapchorwa district, about 300 hectares of wheat were destroyed, tea estates were flooded making tea picking difficult and coffee exports dropped by 60% between October and November (accruing from disrupted transport system) (January 2002 Uganda's Report to UNFCC cited in Ben 2005).

In addition, infiltration of water resources and flooding of some pumping stations (submerging of pumping stations) were enormous in the country, threatening water availability and diseases, such as cholera (ibid). Furthermore, although the monetary value of the losses in the agricultural sector and inaccessible markets was not estimated; the total cost could run into hundreds of millions of US dollars (Source: January 2002 Uganda's Report to UNFCCC cited in Ben 2005).

5.2.3 Experiences of floods in Kenya

Kenya is already affected by extreme climatic events, especially floods, droughts and strong winds (Awuor 2008). These climate-related disasters are projected to increase in frequency and intensity with long-term climate change (ibid). Already, floods are reported to damage productive land, thus causing agricultural losses and increased food insecurity. Floods also damage transport and telecommunications infrastructure such as roads, bridges and pipelines, as well as electricity and telephone lines (Awuor 2008).

The 2004 Indian Ocean tsunami and the 2006/07 flooding experienced on the Kenyan coast exemplify the risks that Mombasa faces. These events led to very large economic losses as major infrastructure and fishing vessels were damaged and one life was reportedly lost in the tsunami. During the flooding in Mombasa in 2006, the Ministry of Health issued a cholera alert and reported 94 suspected cases of cholera on the coast between 20 October and 11 November 2006. Thirteen cases were found to be positive for cholera and at least two deaths were reported. In addition, water sources were contaminated; several drainage systems collapsed and water pipes were washed away. The Kenya Red Cross estimated that approximately 60,000 people were affected by the floods in the coastal part of Kenya, a high proportion of whom were in Mombasa (the coast's main population concentration) (Awuor 2008).

5.3 Adaptation to extreme events

5.3.1 Adaptation experiences from Tanzania

There is a diversity of adaptation experiences in Tanzania, mostly done through traditional knowledge of past climate change events. General observation has shown that those with least resources and power are obviously the least likely to be able to adapt rapidly to change their livelihoods and survive (Nelson & Stathers 2009).

For example, farmers in the semi-arid areas have changed the balance of crops grown due to climate, market and government advice signals, with greater cultivation of drought-tolerant crops, small scale irrigation of crops, the practice of crop rotation, increasing non farm income generating activities and use of appropriate crop varieties (early maturing). Farmers are choosing different faster-maturing sorghum varieties because the rainy season is now so short that their traditional varieties cannot mature in time. Sesame and sunflower have been introduced because they are more drought-tolerant. Cassava production has increased, because it is a drought-tolerant food crop (Morris *et al.* 2006; Nelson & Stathers 2009; URT 2007; Majule 2009). Also, the increasing unpredictability of the rainfall season in the regions has led to more people having to use oxen ploughs. Ploughing land using oxen is much faster than by hand and this speed allows maximum use of the shortened, often intermittent rainy period for crop production. However, the poorest households can rarely afford to plough using oxen and the wealthier owners prepare their own fields first (Nelson & Stathers 2009).

O'Brien *et al.* (2000) found that the greater proportion of farmers from Morogoro and Iringa regions used the following strategies in response to the seasonal weather average rains during the 1997/98 season. The widely used adaptation approaches include switching between crops, altering the mix of crops grown and changing the timing of planting in the light of evidence they obtain of the growing season (O'Brien *et al.* 2000). For example, farmers may switch from maize to sorghum and/or cassava when there is a threat of drought or food insecurity and switch to rice or banana when rainfall is abundant. Crop switching has shown a positive trend in adapting to drought and floods in Tanzania (Paavola 2004; O'Brien *et al.* 2000). Illustrations of adaptation mechanisms are presented in Photos 17, 18 and 19 and the associated case studies.

Photo 17: Cultivation of drought-resistant crops such as cassava is an important response strategy to climate extremes by farmers.

The case of traditional irrigation (vinyungu) farming in Iringa Region

This case study was taken from work done by Majule & Mwalyosi 2006 and Ravnborg 1993

Vinyungu farming is a traditional farming system in Iringa Region practised by smallholder farmers, usually in valley bottoms or flood plains. Ideally, these areas are characteristically moist for a long period of the year. Vinyungu is a local term, which refers to farmlands or fields in valley bottoms or floodplains cultivated during the dry season, utilising natural moisture or water diverted from rivers/streams or harvested from rain to produce food and cash crops. In doing so, farmers to a large extent are coping with the problem of moisture stress common during the dry season. This type of farming is possible in Iringa because the ground water table in most places is relatively high. However, the scale of farming has been progressively increasing, being associated with an increased use of agricultural inputs in order to maximize production due to commercialisation of crops and vegetables. Thus, due to evolution in agricultural practices, traditional vinyungu are being transformed in order to increase productivity. However, experience from this study indicates that there are still areas (such as Mlafu Village) where a minimum level of agricultural input is being used in vinyungu. There is increasing importance of vinyungu farming practices. For this reason, the demand for water for irrigation is also on the increase. Water availability is largely dependent on wetland characteristics, rainfall, temperature regimes, soil physical properties and location of the farm relative to the river or wetland system.

Photo 18: Picture showing traditional systems of preserving water during the dry season

While adaptation strategies in Morogoro and Iringa regions, Dodoma, Tabora and Singida rely on agricultural approaches with minimal off farm activities (Nelson & Stathers 2009; CCV 2009; Mary & Majule 2009; O'Brien *et al.* 2000); Eriksen *et al.* (2005) found in Saweni Village, Same Distri ct in Tanzania that households' coping mechanisms during drought included casual labour, brick making, handicraft, collecting honey and charcoal burning. Indigenous fruits were also highly regarded because they could be harvested by any household member and did well in drought conditions (Orindi & Murray 2005).

These activities provide an important source of cash to allow households to purchase food and cater for other necessities at such times. Remittances from migrant family members and relatives play an important role in household well-being during difficult periods. People who receive remittances tend to be less affected by shocks in terms of access to food, health services and school attendance (Eriksen *et al* 2002; Orindi & Murray 2005).

Photo 19: Irrigation channel at Ilula

Likewise, experience from CARE food security projects in Mwanza shows that in response to recurrent unfavourable rain seasons, pastoral communities are forced to sell part of their herds as a coping mechanism. The effect is felt at the household level, in the form of reduced nutritional intake among household members. In some other communities, men sell charcoal and women are forced to sell firewood. It is valid to say that local communities need help to adapt to climate change, as these coping strategies may lead them to become more vulnerable not only to changes in the natural resource base but to food and nutritional insecurity (CARE 2006). Initial vulnerability assessments and adaptation planning from Tanzania illuminate on the need for mangrove protection, reforestation with "climate-smart species", integrated land-use and marine planning as well as activities to improve resource use technology. Coordinating the testing of adaptation methods in geographically diverse locations within a common habitat type aims to increase the replicability so that project results can be transferred to other conservation efforts around the globe (Hansen *et al.* 2003).

Finally, literature has indicated that women's indigenous agricultural knowledge supports household food security, especially in hard times (for example, drought and famine), where they can use their knowledge of drought and pest-tolerant plants and seed selection to cover diverse conditions in a growing season (IPCC 2007: 457). Recognition of these gendered knowledge systems and skills may provide a rich resource for coping with climate change (Nelson & Stathers 2009:84; Rossi & Lambrou 2008).

5.3.2 Adaptation experiences from Kenya

Observations indicate that impacts of droughts in Kenya have been more enormous and disastrous than floods that concentrate much along the coast. While adaptation to floods require infrastructural mechanisms; in some instances, for example, droughts have been countered by digging and maintaining sand dams in river bottoms. The dams allow for continued cattle watering during dry periods and have reduced cattle deaths and conflict (SEI 2008).

While Tanzanian approaches have been combating droughts through diversified crop and farming practices, much of drought adaptation in Kenya has concentrated much on sustaining livestock keeping. However, this does not imply absence of agricultural adaptations measures in the region; rather it is a matter of scale in operation (Eriksen & Lind 2009). In some areas, some members of the community had their herds completely wiped out due to droughts. This has pushed them to subsistence crop farming. Maize and beans are being planted and some families have had relatively good harvests to last a month or two from the previous season. On the other hand, most families who could not change their lifestyles have found it hard to cope.

Vulnerability of communities in Kenya is further complicated by occupation of marginal degraded lands and low education and skill level (IDRC/CCAA 2009). Surprisingly, the Maasai still resort to traditional adaptation techniques even where modern lines of assistance fail. This involves moving cattle to less vulnerable ecosystems for grazing and water during drought periods (ibid). However, this trend of adaptation has inflicted conflict between farmers and pastoral societies in the Maasai, both in Tanzania and Uganda.

5.3.3 Adaptation Experience from Uganda

National Level Initiatives

Uganda has so far adopted National Adaptation Programmes of Action (NAPA) and developed national inventory of GHGs as one step forward to mitigate Green House Gases (GHG) emissions in her national priorities and aspirations. However, there is limited vulnerability assessment done in sectors such as agriculture, water resources and forestry. Therefore, a lot has been done to communicate urgent and immediate adaptation interventions required to minimize adverse effects of climate change. In addition, Uganda has signed and ratified many climate change international policies. Also, national policies have also been developed and up dated to acclimatize adverse effects of climate change (Ben 2005).

Local Level Initiatives

Local communities have learnt proper utilisation of rainfall patterns in the country, mainly the uni-modal and bimodal patterns. Hence, because rainfall patterns differ from the north to the south in Uganda, these variations are coupled with the changes in the dominant crops and cropping systems. For example, in the north where rainfall is almost uni-modal, annual crops such as millet and other grains are prominent. Late-maturing varieties of sorghum, sesame and pulses are also planted, as are cassava, legumes and other vegetables (Phillips & McIntyre 2000).

Furthermore, grains such as maize are usually planted first in March and then again in August or September so that physiological maturity will occur in the dry season. Perennial crops, such as banana and coffee, dominate most of the cropping systems of the bimodal rainfall zone, although locally distinct cropping systems (e.g. grain in south-western and eastern highlands) do exist. Early-maturing grain crops (for example, 120 days to maize dry harvest) and pulses are grown in both rainy seasons (Phillips & McIntyre, 2000).

Other commonly planted crops include sweet potato, Irish potato, cassava and legumes. Planting dates for annuals vary with the onset of the rains. Depending on soil moisture, maize may be planted from mid-August to mid-September in the first season and mid-February to mid-March in the second season. Beans are often planted well into April and October. Given the dependence of planting time and crop choice on rainfall distribution, there could be potential for utilizing forecasts of season arrival date and duration in crop management (Phillips & McIntyre, 2000).

CONCLUSION

There are numerous arguments supporting that adaptation to climate change impacts is necessary, that it is already occurring and will occur with greater urgency in the future at a range of scales (Adger *et al.*, 2005 & 2007). The implication for adaptation therefore may be to not only cushion adverse impacts, but also to harness positive opportunities. This suggests consideration of an enhanced portfolio of linked-adaptation responses – for example a strategic shift from maize to cash crops over the medium term, and inter-basin transfers in the case of water resources. Such strategic shifts however may entail economic and dislocation costs – and therefore require careful screening, particularly with regard to their effects on equity and rural livelihoods. More rigorous testing of particular crop and stream-flow projections may also be advisable prior to undertaking such adaptation responses (Agrawala *et al.* 2003).

Cropping practices that are often used to mitigate the effects of variable rainfall include: planting mixtures of crops and cultivars adapted to different conditions as formal or informal intercrops; using crop landraces that are more resistant to climate stresses; using crop trash as mulch; planting starvation-reserve crops; and a variety of low-cost water-saving measures (Challinor *et al.*, 2007). Other agricultural strategies include management of tillage practices, for example, minimum or no tillage in hazard areas, planting cover crops and applying green manure to help restore soil fertility where leaching occurs from increased rainfall. Many farming communities grow more than one crop as a form of insurance against total crop failure (Orindi & Murray, 2005; Majule 2009). Such coping responses at the farm-level can become insufficient when droughts are more widespread and severe, particularly when consecutive drought years lead to loss of seed stocks and biodiversity and/or draught animals or are combined with low capital reserves for coping and with other economic or social stresses to the food system.

Thus, farmers can cope up to a certain limit and their livelihoods can maintain a measure of resilience to shocks, but not indefinitely. Once their capital assets (e.g. savings, seed stocks, drought animals, social capital) erode away beyond a certain threshold level, they are forced to succumb in the absence of any effective local or national level support mechanism such as for replenishing seed stocks or draught power or non-farm employment (Bird & Shepherd 2003; Challinor *et al.* 2007).

Therefore, a key ingredient in the ability of farmers to cope with or adapt to climate variability and change is their access to relevant knowledge and information that will allow them to modify their production systems. Some of this knowledge is already part of local knowledge systems, such as varying planting dates in response to seasonal variations in rainfall onset and intensity; some will come from outside the local system, such as new varieties more tolerant to drought or with shorter growing seasons. Current and prospective institutional changes in the way knowledge is created and information communicated offer grounds for cautious optimism that the availability of and access to appropriate knowledge will improve (Challinor *et al.* 2007).

CHAPTER SIX

SOUTHERN AFRICA SMALLHOLDER FARMERS' EXPERIENCES
IN A VARIABLE AND CHANGING CLIMATE
Case Studies from Zimbabwe and Zambia

6.0 Introduction

Zimbabwe is now a warmer country than it was at the beginning of the twentieth century (Mano & Nhemachena 2006). The annual-mean temperature has increased by about 0.4°C since 1900, and the 1990s decade has been the warmest in this century. This warming has been greatest during the dry season. During the wet season, day-time temperatures have warmed more than night-time temperatures (Hulme & Sheard 1999). That daytime temperatures over Zimbabwe have risen by up to 0.8°C from 1933 to 1993, which translates to a 0.1°C rise per decade (Unganai 1996). Zimbabwe is expected to warm somewhat more rapidly in the future than the global average. In model scenarios, annual warming reaches about 0.15°C per decade under the B1-low scenario, but this rate of warming increases to about 0.55°C per decade for the A2-high scenario. Moreover, rates of warming are expected to be slightly greater than this during the dry season and slightly less during the wet season. By 2050 temperatures and rainfall over the country will be 2–4°C higher and 10–20% less than the 1961-90 baselines respectively (Zimbabwe Initial National Communication [ZINC] 1998).

There has been an overall decline of nearly 5% in rainfall across Zimbabwe during the 20th century, although there have also been substantial periods - for example, the 1920s, 1950s and 1970s - that have been much wetter than average. The early 1990s witnessed probably the driest period in the 20th century, a drought almost certainly related to the prolonged El Niño conditions that prevailed during these years in the Pacific Ocean (Hulme & Sheard 1999). Zimbabwe was characterized by low precipitation during the late 1920s to 1949, late 1950 to about 1972 and from 1980 to present (Unganai 1996). The decade 1986-1995 in Zimbabwe was about 15% drier than average (Hulme & Sheard 1999). In terms of precipitation in Zimbabwe, of the 14 years from 1990/1991 to 2003/2004, at least ten years in each agro-ecological zone had below normal rainfall (Gandure & Marongwe 2006). Model experiments predict annual rainfall decreases across Zimbabwe in the future. This decrease will occur in all seasons, but predictions are more conclusive for the early and late rains than for the main rainy season months of December to February. Furthermore, by the 2080s, there will most likely be annual rainfall averages between 5% (B1-low scenario) and 18 % (A2-high scenario) less than the 1961-90 average.

The ENSO is one of the main causes of climate variability for many tropical regions, especially for Zimbabwe. For example, since about 1976, there has been a tendency for negative (El Niño) warm phases of ENSO to dominate. This period has seen very strong El Niños in 1982/83 and 1997/98 and a prolonged El Niño between 1991 and 1994, events which are considered to have been partly caused by global warming (Hulme & Sheard 1999).

For Zambia, the observed temperature from 32 meteorological stations in the country was analyzed to detect trends in temperature change over last 30 years. The mean temperatures computed for the agro-ecological zones for three time periods, November–December, January–February and March–April, indicate that the summer temperature in Zambia is increasing at the rate of about 0.6°C per decade, which is ten times higher than the global or Southern African rate of increase of temperature (Chigwada 2004; de Wit 2006; Hulme 1996; Jain 2006). The rate of increase is highest in November–December as compared to other periods across all agro-ecological zones.

The rainfall anomalies from the 1970–2000 annual averages were computed using observed data from all 32 meteorological stations in Zambia for the agro-ecological zones. These annual rainfall anomalies indicate that of the 14 years from 1990/1991 to 2003/2004, at least ten years in each agro-ecological zone had below normal rainfall. We further note that the variability in annual totals across the three agro-ecological zones has not been uniform. The southern zone (Zone I) has experienced more severe dry seasons than the central zone (Zone II) in the last 20 years. Moreover, Zambia has had some of its worst droughts and floods in the last two decades (De Wit 2006). Southern Zambia experienced severe floods in the period 2007 to 2008. In a study done in Southern Zambia by Kurji *et al.* 2006, with a focus on the exploratory analysis of daily rainfall data for evidence of climate change, while there is an indication of a slight trend to reduced annual and seasonal rainfall over the last 30 years, there is no obvious indication of more variation in the recent years than in the past. In this regard, the climate change has resulted in more variability, rather than a change in the mean. Moreover, while there is a slight trend to lower total seasonal rainfall and a lower number of rainy days over the years; the average amount of rainfall per rainy day, on the other hand, does not seem to have been affected.

As is the case in Zimbabwe, the ENSO phenomenon is now recognised as the major factor in determining precipitation patterns in Zambia especially during the summer rainfall, October to April. ENSO affects International

Tropical Convergence Zone (ITCZ) and Congo Air Boundary (CABS), the main rain bearing mechanisms. The opposite phenomenon, La Nina, is considered to bring more rainfall, which normally results in floods. In the same respect, the ITCZ phenomenon is contrasted by the Botswana Upper High Influence (BUHI) which controls drought episodes and uneven rainfall distribution. BUHI creates an unfavourable condition for rainfall by pushing the rain-bearing ITCZ and active westerly cloud bands out of the region and Zambia (Chigwada 2004).

6.1 Climate change impacts in Zimbabwe and Zambia

6.1.1 Climate change impacts on agriculture

There has been extensive research on the impacts of climate change, but little has been done on the impacts on agriculture in Zimbabwe (Mano & Nhemachena 2006). This provides a context for this study to investigate the impacts of climate change on agriculture in Zimbabwe, considering that agriculture remains the backbone of the country's economy. The agricultural sector contributes about 17% to the country's GDP (FAO 2005). Agriculture is also an important source of raw materials, providing about 60% of raw materials for the manufacturing sector in the country (Bautista *et al.* 2002; Poulton *et al.* 2002). For instance, drought years that are depicted by negative rainfall deviation correspond with the declining and low growth rate in GDP contribution from the agricultural sector, implying that rainfall patterns have a significant effect on this contribution over the years. Notable examples are the growth rates in the early 1970s, early 1980s and the 1992 drought. More droughts were witnessed in the seasons 2002/3 and 2007/2008. During these drought years the temperature increased and the rainfall was very low, and this had a significant effect on agricultural performance and hence the growth rate of GDP contribution from the sector (Mano & Nhemachena 2006).

Most crops that are highly sensitive to climate and ecosystems will shift over space in response to climate change. For instance, research done in various countries in Southern Africa has demonstrated that a 2°C rise in ambient temperature and a rise of mean temperature by 4°C have significantly lowered yields. Potential effects of climate change on corn, using a GCM and the dynamic crop growth model CERES - maize in Zimbabwe, showed that maize production was expected to significantly decrease by approximately 11–17%, under conditions of both irrigation and non-irrigation (Agoumi 2003; Magadza 1994; Makadho 1996; Mano & Nhemachena 2006; Muchena 1994; Stige *et al.* 2006). In this regard, it is suggested that major changes in farming systems can compensate for some yield decreases under climate change, but additional fertilizer, seed supplies, and irrigation will involve an extra cost. In addition, results of analysis of potential impacts using dynamic simulation and geographic databases reaffirm the dependence of production and crop yield on intra-seasonal and inter-annual variation of rainfall for South Africa and the Southern Africa region (see Hulme 1996; Schulze *et al.* 1995; Schulze 2000).

A study done to see how agricultural production would respond to climate change in Zimbabwe indicates that a 2.5°C increase in temperature would decrease net farm revenues by US$ 0.4 billion. A 5°C increase in temperature would decrease net revenues across all farms, dryland farms and farms with irrigation by USD 0.4 billion, USD 0.5 billion and USD 0.003 billion respectively. Moreover, a 7% and a 14% decrease in precipitation would result in a decrease in net farm revenue by USD 0.3 billion for all farms (Mano & Nhemachena 2006). Irrigated farms are therefore portrayed to be more resilient than farms under rain-fed agriculture. Even though irrigation will boost maize production in all areas, the yields are lower under climate change conditions than under normal climate. The reduction in mean seasonal precipitation under climate change conditions implies that the water available for irrigation purposes would also be affected accordingly and this will reduce the effectiveness of irrigation as a strategy to combat the effects of climate change. More importantly, broad-scale shifts in agricultural capability due to climate change would affect rural livelihood and the national economy. Subsequently, all vulnerable groups are threatened by climate change through the ripple effects that diminish the resource base and increase the possibility of resource conflicts and tensions between the agricultural and industrial sectors (Matarira *et al.* 1995).

Zimbabwe has witnessed discernible shifts in agro-ecological[15] regions (see Figure 1). With higher temperatures projected to shorten the growing season of crops by two to thirty-five days, one of the obvious consequences is reduced yields as well as decreased livestock. In the study done by Matarira *et al.* 1995, maize production at all stations is more consistent under normal climate than under climate change conditions. Climate change introduces greater variability in maize yields, thus making maize production a more risky agricultural activity. This will impact negatively on the socio-economic lives of the people in general and food security in particular. For instance in Masvingo, which represents Natural Region IV in this study there is a strong likelihood that climate change will make the Region a non-maize producing area. Implications herein are

---

15 Zimbabwe is divided into five agro-ecological regions in a continuum, with region one receiving the highest rainfall and region five the least.

that Natural Region IV, which represents 42 % of communal areas, will not adequately supply its population with the staple food crop. This is contrary to the MDG 1, which happens to be central and related to all the others and spells out the need to eradicate extreme poverty and hunger.

Natural Climatic Regions of Zimbabwe

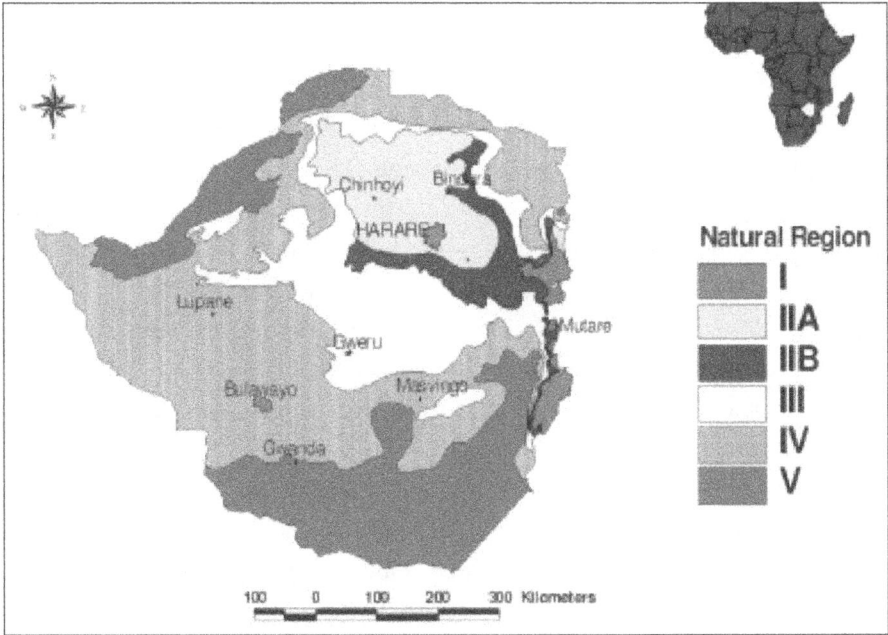

Figure 1: Agro-ecological zones in Zimbabwe (Source: FAO Sub-Regional, 2005)

Because Zambia's economy is agriculture based, drought is likely to have an adverse effect on food security, resulting in lower economic growth. It is anticipated that there will be reduced precipitation; high temperatures and evapo-transpiration during droughts, which will affect the staple food production. Similarly, floods and droughts will affect agricultural production, which can be worsened by the occurrence of pests. On the other hand, removal of vegetation cover through fires or overgrazing increases the risk of soil erosion during heavy downpours. This, too, leads to decreased agricultural productivity (Chigwada 2004). For example, excessive rain in 2001 and dry spells during the 2001/02 growing season led to a major shortfall in maize production, a decrease of 42% compared with the average yearly production. Compounding the plight of agriculture in Zambia is the incidence of HIV/ AIDS due to unavailability of labour as the agricultural production among smallholder farmers is labour-intensive (Chigwada 2004).

Agriculture is becoming an increasingly important sector in the Zambian economy since the mineral sector, which was the backbone of the economy from post-independence times (1964) till the late 1980s, has declined. The agriculture sector generates about 18% to 20% of the country's GDP, provides a livelihood for more than 60% of the population and employs about two-thirds of the labour force (Jain 2006). Agriculture in Zambia depends on rainfall to a very large extent. In this regard, the negative impacts of extreme climate events which are believed to be manifestations of long term climate change have affected crop production in the country since the 1990s (Jain 2006). The last two decades have seen Zambia experiencing some of its worst droughts and floods, notably, the droughts of 1991/92, 1994/95 and 1997/98 worsened the quality of life for vulnerable groups such as subsistence farmers (Muchinda 2001). For instance, the yield during the severe drought of 1991/92 was less than half that of 1990/91. In the seasons 1972/73, 1979/80, 1981/82, 1983/84, 1986/87, 1993/94 and 1994/95, significant shortfalls in maize yield were also recorded and these seasons were characterized as having below normal rainfall by the Zambian meteorological department. Essentially, drought has been the biggest shock to food security in the country during the last two decades (MoA 2000; Muchinda 2001).

In addition, the impact of extreme climate events has been felt in substantial loss of livestock and fertile soil. Low productivity in the agricultural sector has contributed to a low GDP. In a study done in Zambia by Jain 2006, it was found that there may be a negative effect if temperature rises by at the beginning of the cropping season when plants are germinating as evidenced by the marginal net revenue per hectare USD322.628 for an increase of 1°C in the mean temperature in November and December. In addition, although the marginal net revenue per hectare of USD315.70 for an increase of 1°C in the mean temperature of January and February indicates that if the temperature rises during the growing stage of the plant, this may have a positive effect on the crop, a decrease of about 20% in the precipitation for this period can reduce the net revenue by about US$334.67 (Jain 2006). Given these findings, agricultural production in Zambia is subject to the uncertainties of extreme climate events, which are indicative of an increasing mean temperature and reduction in total seasonal rain on a long-term time scale.

6.1.2 Climate change impacts on water in Zimbabwe and Zambia

In Zimbabwe, climate change is considered to have impacts on water supply, water demand and on water conservation reliability. A doubling of CO_2 would cause the rivers in the Eastern Highlands of Zimbabwe that are well watered and perennial to develop flow regimes similar to those currently experienced in the dry regions i.e. seasonal rivers. It is further projected that climate change will increase irrigation water requirements due to increased potential evapo-transpiration for the doubling of CO_2 scenario (ZINC 1998). Catchments will be water scarce as a result of challenges such as increase in demand due to population growth and allied uses, which will be compounded by climatic change.

The major form of water conservation in Zimbabwe is through the construction of dams. All urban centres and large-scale irrigation schemes depend on dams for water supply. The vulnerability of Zimbabwe to climate change is indicated largely by the impact of climate change on future water supply from dams. The amount of water that can be supplied by these dams with a reliability of 96% was estimated for the baseline and doubling of CO_2 scenarios. The costs of constructing reservoirs with such storage capacities were estimated to be extremely high. The analysis further shows that the yield of dams will decrease by about 30%-40%. If the same level of supply and reliability is desired, then there will be a need to either increase the storage capacities of these dams or construct new ones. Increase in these capacities is not possible since all major dams are designed for their maximum yield (ZINC 1998). With regards to Lake Kariba, generating capacity would decrease by as much as 50% during dry periods. And the maximum generating capacity would barely exceed 50% of installed capacity during wet years (Arnell 2004; Urbiztondo 1992).

In addition, model simulations show annual rainfall for Zimbabwe declining by 5 – 20% of the 1961-90 average by 2080 in all the country's major river basins. Agriculture has been identified as the sector most vulnerable to these climatic changes (ZINC 1998). Estimations of water demand to year 2075 were based on population projections and average growth rates in water usage from 1950 to 1995. Rainfall-runoff simulation for the doubling of CO_2 scenario showed that a 15%-19% decrease in rainfall and a 7.5%-13% increase in potential evapo-transpiration will result in a 50% decrease in runoff in the Gwayi, Odzi and Sebakwe catchments. Therefore, the climate change impact on runoff among the three representative catchments considered was a 50% decrease.

El Niño affects the precipitation in Zambia resulting in drought while La Nina is associated with floods. In the Kafue Basin there are already conflicts on water rights due to water demand for various uses including agriculture, hydro-power generation, industry and domestic demand, which are considered to intensify with climate change. Zambia has a highly vascularised river system along with water bodies that cover as much at 6% of the total land area, which is dominated by the Zambezi River drainage system. Growing demands for water for hydroelectric power, urban agriculture and industrial consumption have led to conflicts over water rights in the Kafue Basin. A study by Arnell (2006b) states that the greatest reduction in run-off by 2050 in Africa will be in the Southern African region, suggesting that water use to resource ratio changes will put countries in the high water stress category. In addition, the shortfall in rainfall and the effect of increases in potential evaporation and the resultant reduction in run-off in the major river basins, including Zambia, could be as high as 40% (Chigwada 2004).

In the same respect, low soil moisture and high evapo-transpiration will promote desertification due to a reduction in vegetation cover. This will lead to soil erosion and sediment discharge that could cause siltation of reservoirs. Drought occurrences are also known to have caused a reduction in wildlife through deaths in the past besides a decline in tourism and reduction in water flows at the Victoria Falls (Chigwada 2004).

6.1.3 Climate change impacts on the economy in Zimbabwe and Zambia

Future decreases in rainfall will have implications for the contribution made by Lake Kariba to the Zambian and Zimbabwean economies (Benson & Clay 1998). Lake levels are crucial for energy generation at the Kariba Dam and also for the animals of the Kariba National Park sited along the banks of the reservoir. This is the case since tourism at Kariba makes a major contribution to Zimbabwe as a source of foreign exchange. The dry years of the mid-1980s led to a major fall in lakes level and subsequently a reduction in energy generation. Scenarios suggest further decreases in rainfall across the Zambezi basin, a trend that would undoubtedly lead to lower lake levels for Kariba. Furthermore, the 1991-92 drought, which hit most of Southern Africa forced the Zimbabwe stock market to decline by 62%, causing the International Finance Corporation (IFC) to describe the country as the worst performer out of 54 world stock markets. The country's manufacturing sector declined by 9.3% in 1992. In the same respect, drought caused a 25% reduction in the volume of manufacturing output, and a 6% reduction in foreign currency receipts (UNEP 2002a). In Zambia, economic impacts from curtailment of hydro-power generation from Lake Kariba on the Zambezi River as a result of the 1991-92 drought were estimated at US$102 million in GDP, US$36 million in lost export earnings and 3,000 job losses (Chigwada 2004). Moreover, because Zambia's economy is agriculture based, drought is likely to have an adverse effect on the agriculture sector, resulting in lower economic growth.

6.2 Case studies

Case study research, on which in-depth case studies presented in this chapter are based, was conducted in Lupane and Lower Gweru districts in Zimbabwe and Monze and Sinazongwe Districts in Southern Zambia between August 2007 and June 2010[16]. The objective of these case studies included establishing farmers' perceptions of climate change and variability, impacts from these changes and strategies that farmers employ to respond to climate change and variability. In-depth interviews were carried out with the sampled farmers, who were part of the bigger action research project in the study sites.

6.2.1 Perceptions of farmers regarding climate variability and change and their causes (case studies)[17]

This section presents in-depth case studies illustrating farmer perceptions. These case studies are of farmers from Lupane and Lower Gweru in Zimbabwe and Monze in Zambia. Photos 20, 21 and 22 illustrate effects of floods and excessive rains in Zimbabwean case studies.

The case of Ellen Sibanda

Ellen Sibanda (79 years old) resides in Mathonsi village in Nyama ward, Lower Gweru. She indicated that over the past five years, rains have become unpredictable as they no longer start at the expected time around October, but rather in November. Within the same period, temperatures have changed and they have also started experiencing prolonged winters into September. She further highlighted that these changes have been caused by the exodus of people to churches and the abandoning of coordinated traditional rituals, which were an integral component of the way of life of these people in Lower Gweru. Traditional rites that were conducted before have ceased. They used to go to Matonjeni (a traditional sacred grove in the bush specifically preserved for traditional rites) to ask for rains from God through their ancestors. Ellen said, 'We used to play drums at Matonjeni to thank God for looking after us and for providing rains, no matter how little. We also at the same time asked for more rains in the coming season.

16 This study was part of a bigger project 'Building Adaptive Capacity to Deal with Vulnerability due to Climate Change', which was funded by IDRC through the CCAA and led by Midlands State University in Zimbabwe in collaboration with ICRISAT and CIAT both in Zimbabwe and ZARI and ZamMet both in Zambia.
17 In-depth interviews were conducted with the household heads (January 2009) for these and other in-depth case studies highlighted in this chapter.

This is why we had good rains in the past. Now, the church has taken over this role and now, no appeasement is granted to our ancestors'. Ellen therefore suggested that God is angry and has decided to punish people; hence these negative changes in climate. Asked if there were any other reasons for climate changes she said that she thought that something had gone wrong in the oceans somewhere but still attributed this to the wrath of God and ancestral spirits.

The case of Busisiwe Mbangwa

Busisiwe Mbangwa (34 years old), who is a widow and resides in Lupane, highlighted that rains have become unpredictable in the past six years. She emphasised the high incidence of excessive rains which she said was a recent phenomenon dating back to the last three seasons. Busisiwe also indicated that there has been an increase in dry spells over the years since around 1998 and that there is now a high incidence of winters which prolong until September, a change which she had noted for approximately the previous consecutive four years. When asked what she thought was causing these changes in climate, she indicated that God was angry and was punishing Zimbabweans for the political unrest that had been in the country for a number of years. She even cited what she called 'endless talks' (*kutaurirana kusingapere)* that were going on between political leaders in Zimbabwe at that time and gave this as an example of the cause of God's anger. Busisiwe remained pessimistic about the future, suggesting that *'If nothing is done about the current political unrest in this country, then we will continue to experience bad seasons as God is punishing us'.*

The case of Jobert Muzyamba

Jobert Muzyamba lives in Muzyamba village in Monze (45 years old) . He indicated that he had started noticing a change in precipitation around 1992, before which there was predictability of the rains, characterised by good seasons. He highlighted that rains were now less predictable and characterised by a late onset. Jobert reported that before 1992, the rains came in October but now they do not receive the first rains until it is mid-November. Although he indicated that he had not witnessed changes in temperature, he had noted excessive rains and floods in 1989 and 2008 respectively. These bad experiences with floods were compounded for him and his neighbour by the fact that they reside within the banks of Magoye River. Jobert emphasised the increase in climate variability as having started around the same time that President Chiluba got into power. In his own words, Jobert said, 'President Chiluba is the one who came with the high variability in precipitation. Before his leadership, this variability was minimal and we had plenty of food from our fields.' Jobert associated this ascendancy of Chiluba to power with the beginning of changes and variability in climate.

General perceptions are that climate is changing. They highlighted specifically how rainfall has become unpredictable and erratic in the past two decades. These changes and variability are characterised by more frequent dry spells, late onset and early ending of seasons, suggesting shorter growing seasons. Perceptions also indicate that temperatures have become warmer, specifically with prolonged temperatures, which have also become warmer than before. Droughts and a recent phenomenon of floods and excessive rains were also highlighted.

The case studies cited in this section suggest that farmers do not only associate changes in climate with natural factors, but also with social and spiritual factors. The implication is that when there are political, social and economic problems in a country, farmers tend to link them to climate change. Essentially, the cultural context and spiritual world view play a critical role in shaping farmers' perceptions and attitudes, a factor which may cloud farmers' consciousness of the negative effects of human activities on the earth. Farmers in Lower Gweru and Lupane link the political crisis in Zimbabwe at this time and the decline of social and cultural practices to the variability in climate.

Similarly, a farmer in Monze associates the beginning of climate variability with the ascendancy of Chiluba into power. The period of his leadership period was marred with controversy and linked to economic problems in Zambia during this period. Moreover, what is emerging is the idea that we cannot disassociate climate change from the political, social (including the cultural and spiritual realms) and economic context. Farmers try to make sense of what is happening in their environment based on the socio-cultural framework in which they operate. However, it is of concern that farmers fail to associate climate variability and change with human activities and rather blame this variability on ancestors. This concern is based on the assumption that if farmers are aware of the extent to which activities such as deforestation may alter the natural processes, these farmers may consider taking remedial action.

6.2.2 Farmers experiences in the face of climate variability and change

This section presents case studies that illustrate farmers' experiences during climate induced floods/excessive rains and droughts in Monze (Zambia) and Lower Gweru (Zimbabwe) respectively.

The case of Nehemiya Hambuya

Nehemiya Hambuya, who is 47 years old, is married and has nine children. He stays in Mujika village, Monze and this is also where he was born. Nehemiya started farming with his parents at the age of 19 but started to farm on his own in 1995 when he was 33 years old. Although he does not know the total amount of land that he holds, his land is not much and he used to rent land when the rains were more predictable and he had adequate draught power. Around 1991, he used to get on average, yields as much as 260 x 90kg bags of maize. From around 1996, peasants in his village started experiencing more and more dry spells and drought periods and his yields dropped to an average of 35 x 50kg bags of maize on the same amount of land as before. These drought conditions also affected his livestock which were attacked by corridor diseases and teaks, subsequently affecting availability of draught power. He lost 20 cattle in a period of 10 years and now has no cattle, and only one goat. Nehemiya attributed this loss of cattle to corridor disease and teaks which were caused by the diminishing dipping facilities and availability of vaccinations. After this loss, he then decreased the hectare under cultivation and stopped renting land. The unavailability of draught power compounded the drought conditions, leading to food insecurity in his home, considering that he has a large family. Making the situation worse for Nehemiya was the reduction of the loan facility for inputs in 1999. The recent incidence of floods in Monze, according to Nehemiya, ushered in a further drastic reduction in crop yield. The plight of his family has worsened in the previous two seasons (2006/7, 2007/8) due to floods. For example, in 2007/8, he got no bags of maize after his crops did not germinate and those that did were stunted. The little maize that they could get they harvested as green mealies in order to feed the starving large family. Trees around his homestead were uprooted by storms and it was difficult to walk around the home because of the floods which covered part of his home (see Photos 1 and 2). The roof of Nehemiya's main house was blown away during this time. Nehemiya indicated that '*although there were positive impacts from these floods, such as the availability of water for domestic use and livestock watering and the availability of fish around the homestead as water was over-flowing from rivers, negative impacts outweighed the positive ones by far. My household has been severely destabilised by these drought and flood occurrences so much that now I find it difficult to plan for agricultural activities for the next season, as before. I am also finding it increasingly difficult to feed my family.*'

Photo 20: The state of Nehemiya's homested during floods (Zambia).

The case of Mollie Dube

Mollie Dube of Nyama ward, Lower Gweru, was born in 1946 and has been a widow for nine years now since 2002. Mollie has six grown up children, some of whom are working in various towns around Zimbabwe and others in neighbouring countries. Her husband worked for the Ministry of Information in Gweru while she stayed in the rural areas all the time. Mollie's husband stopped working around 1998 when he fell ill and came back to the rural home to stay with his wife. At this point he was now farming with her. She attended school up to standard six where she says she learnt 'proper Oxford English'. Both Mollie and her husband are originally from Lower Gweru so they both married locally. Mollie has experienced food insecurity due to excessive rains in the 2008/2009 season (see Photos 3 and 4) and droughts in the past years. Crop cultivation was her major source of income and this has meant that her income has gone down drastically. She used to sell her harvest to the Grain Marketing Board (GMB) but she has had to stop. For 10 years now she has not sold her harvest to the GMB. For her, problems brought about by droughts and excessive rains have been compounded by the death of her husband, who would finance agricultural activities in the farm and address the issue of lack of inputs. She no longer has enough money to buy inputs as her major source of income has been negatively affected. She further highlighted that commercial farmers who used to produce seed are no longer there and when seed is available it is extremely expensive, a factor which has been made worse by hyper-inflation in Zimbabwe. In addition, Mollie has become more vulnerable since the death of her husband as she has started to have her agricultural implements stolen as people are aware that she has lost her 'protector' (her husband). Although she owns seven hectares of land, she is only cultivating one hectare due to this combination of constraints. Mollie can no longer afford to hire labour for her large piece of land as before. Ever since this time, she has been growing crops only for subsistence but has been failing to get any yield to last her until the season.

Photo 21: Water logging at Mollie's homestead after excessive rains

Photo 22: Mollie's stunted crop maize after excessive rains

The case studies highlighted in this section emphasise that in addition to climate variability and changes, farmers are faced with a myriad of other challenges, which worsen their predicament. Therefore, it would be naïve to attribute food insecurity solely to climate change, a factor which suggests that there is a need to understand the multiplicity of challenges that confront farmers, in addition to climate change.

6.2.3 Responses to climate variability and change among farmers in the study areas

This section illustrates farmers' concrete experiences in responding to climate variability by presenting case studies from Lupane (Zimbabwe) and Monze (Zambia) districts.

The case of Jophina Sabe

Jophina Sabe from Lupane district is chronically ill and was born in 1957. She has been widowed since 1998, after being married for 18 years. After her husband's death, Jophina's husband's relatives took away all her property. She had to relocate back to her parents' home. Jophina has three children, two of whom are doing menial jobs in the city and one is living with her. A number of factors have led to her problems as in addition to not having livestock for plough power and the problems in the country, she feels that the erraticity of the rains has really affected her household. Of late, it has been difficult, if not impossible, to get agricultural inputs on the formal market and when they do get them it is late in the season, around December, and the prices are prohibitive. Back in the 1980s, Jophina indicates that she would get maize yields as high as 104 x 90kg bags. However, those times have passed and in 2004/5 she got 5 x 50kg bags and in 2007/8 she had no bags due to excessive rains. Her fields were washed away and she lost all her crop. Jophina highlighted that when yields started going down with bad seasons from around 2001/2; she decided to engage in petty trade and has a market stall. 'When I realised that my family was about to starve due to crop failure, I decided to go to Gweru (around 2004) and partner with a friend. We hired a market stall to sell our commodities from. However, this business did not last long as we became broke after we were heavily affected by escalating inflation around 2006 going to 2007. We no longer got much from the business in terms of profits, which is why I abandoned the business and returned to the village but by then my son was already in Form 3, after which he continued with night school.' Jophina has a small garden which does not yield much except for subsistence. She had therefore resorted to buying produce from other farmers' gardens and selling it to get profit. She would sell the produce to gold panners, but when her sales were again eroded by inflation she stopped this trading. This was her major livelihood strategy, thus she has been affected heavily as her income has declined. As a result, she now relies on receiving aid from CARE. She receives a 20kg bucket of maize, one litre of cooking oil and beans. This allocation usually lasts for three weeks, after which she eats vegetables only until she gets the next allocation. Jophina, at the beginning of each season, puts compost in her maize crop. She picks manure from gardens and pastures and also collects leaf litter to make compost for her crop. Of late she has started growing maize alone as she cannot get seed to diversify, since 1998. While she owns only one hectare of land, Jophina does not recall a season where she cultivated the whole piece of land. In 2007/8, she even cultivated one acre as this was the only area on which she had done winter ploughing. After the first stage of winter ploughing, there were excessive rains which did not allow her to increase the area under cultivation. She recalled that prior to 2000, she would do winter ploughing on the whole piece of land.

The case of Maria Mweemba

Maria Mweemba was born in 1937 and is the first of five wives in a polygamous marriage to the headman of Sikaula village in Monze. Between them, they have 35 children but almost all of them have left either to stay in cities or to have their own homesteads in the same village once they have married. Maria indicated that her family has in the past been affected by droughts and recently, floods. She gave an example of the most recent dry years that she could recall as 2005/6 and 2006/7. Thereafter, the next season that followed (2007/8) was characterised by floods. In these dry years, she recalled that their crops dried up and their maize crop was affected by pests from the root and others from the heart. Their wells in their home and gardens also dried up during these periods. In the flood season, there was water logging in their fields and even had water reaching knee height in the fields. In addition, there were so many malaria cases at her home with the grandchildren mainly being the worst affected. While as a homestead they would get an average of 520 x 50kg bags of maize yield under normal circumstances, in this period they got only 50 x 50kg bags of maize as the overall yield. For the family, all the wives work in the fields collectively and have one storage place and share the yield. Maria indicated that before these crises, her family never had to join cooperatives for loan schemes as they were always self-sustaining and could afford to purchase agricultural inputs for the following season. They even over the years managed to acquire sound agricultural implements that include tractors, harrows, cultivators and ploughs, among others (see Photos 23 and 24). Although she failed to state the exact number of livestock that the household has, she indicated that there are plenty which belong to the husband and separate ones for each wife, herself having seven cattle and nine goats. In recent years, the case is different as they will have got low yields in the previous season, with food not being enough, they would therefore have to buy more in order to feed the large family. In the previous season, they joined a cooperative for acquiring inputs and even then, what they acquired was not enough. To deal with these food shortages, Maria's family has had to resort to buying food over the years since the droughts started. Also in drought periods, the wives intensify their gardening in addition to crop cultivation from which they expect to yield less when the season

Photo 23: Agricultural implements owned by Maria's household (Zambia)

is bad. Essentially, the major source of livelihood for the family is crop cultivation but they intensify gardening activities when they get low yields. Moreover, remittances from their children have gone a long way in cushioning them in the years that they have not had good yields. They also decreased the area under cultivation in this season although she could not talk in terms of the exact hectarage.

Photo 24: Tractor owned by Maria's household (Zambia)

Response strategies cited by farmers in the case studies in this section point to the need to enhance strategies that help farmers deal with problems arising from both climate change induced droughts and floods and other multiple stressors that confront these farmers. For instance, the hiring of labour and having a large family is seen to be very important and as a way of adapting to food shortages in the household, a belief which is relatively common in Monze and Sinazongwe (indeed a survey carried out in the study areas under the same project support this assertion by showing the significantly higher percentages of polygamous households in Zambia districts than there are in Zimbabwe districts). In essence, it is important to understand the prevailing conditions in a specific area in order to understand and enhance livelihood activities which may not be found in a different area. Invariably, adaptation to climate changes and variability is local. The family in Zambia had over the years managed to remain self sufficient with the availability of labour. In addition, the availability of draught power in the household is critical in enhancing yields as reflected in the case studies cited. The case of Jophina also emphasises the fact that livelihood diversification from agricultural activities has been a critical strategy that is employed by farmers to supplement livelihood. However, it is important to note that farmers still engage in agriculture based strategies in addition to diversification into non-farm activities.

CONCLUSIONS

There was concurrence among farmers in all sampled districts that there has been a shift in the onset of the rains from October to mid and sometimes late November. In addition, temperatures have risen significantly over the years and winters, which now prolong, have also become warmer than before. Therefore, this study found that farmers have the capability to perceive changes and variability in local climatic conditions. Farmer reported trends in climate variables, which are congruent with what has been observed by meteorologists as cited in literature. This ability to perceive changes in climate may go a long way in enabling farmers to respond to these changes. While climate variability and change remain the most critical and exacerbate livelihood insecurity for farmers who may be already vulnerable, there are multiple stressors that confront farmers. This chapter therefore recommends that studies on impacts of climate change on smallholder farmers need to emphasise on addressing climate change impacts not as an isolated priority but in conjunction with other equally pressing social, economic, and environmental issues.

CHAPTER SEVEN

CROP PRODUCTION ADAPTATION TO CLIMATE CHANGE
AND VARIABILITY
Experience from East Africa

7.0 Introduction:

An Overview of climate change and crop production

Africa is the most vulnerable continent to climate change impacts affecting livelihoods of the majority of people (Lobell & Burke 2008; Challinor *et al.* 2007; Wheeler *et al.* 2000; Stern & Easterling 1999). Therefore, increased variation and changes in mean temperature and precipitation are expected to dominate future changes in climate as they affect crop production (Reilly *et al.* 1995; Porter & Semenov 2005; Burke *et al.* 2009; Lobell *et al.* 2009) via linear and non-linear responses to weather variables and the exceedance of well defined crop thresholds, particularly temperature (Porter & Semenov 2005). These projected consequences are likely to be significant in the Eastern zone of Africa where the majority depends on agriculture for their survival (Challinor *et al.* 2007).

Available literature shows that small-scale farmers who are dependent on low-input and low-output rain-fed mixed farming with traditional technologies dominate the agricultural sector in Eastern Africa region (IPCC 2001; Cooper 2004; Deressa & Hassan 2009). For instance, Cooper (2004) established an estimate that 89% of cereals in sub-Saharan Africa are rainfed. In addition, the World Bank (2002) states that agriculture is clearly the most important sector of the Tanzanian economy which contributed to approximately 45.1% of Gross Domestic Product in 2000. It is also estimated that about 80% of the population of the country relies directly on agriculture of one sort or another for their livelihood (ibid).

Agriculture in Africa is highly dependent on climate (Salinger *et al.* 2005; Boko *et al.* 2007). It is projected that overall crop yields in Africa and East Africa in particular may fall by 10–20% around 2050 because of warming and drying, but there are places where yield losses may be much more severe, as well as areas where crop yields may increase (Jones & Thornton 2003).

Accordingly, it is established that predicted changes in climate will have significant impacts on Tanzania's rain-fed agriculture and food production. For instance, warming will shorten the growing season and reduce rainfall and water availability. Warmer climate will accelerate crop losses due to associated warmer effects from weeds, diseases and pests. In addition, regional predictions indicate that Tanzania may suffer a loss of over 10% of its grain production by the year 2080 (Parry *et al.* 1999: 62-64; Downing 2002). Besides, many developing countries in Africa are seen as being highly vulnerable to climate variability and change (Slingo *et al.* 2005); in part because they have only limited capacity to adapt to changing circumstances (Thomas & Twyman 2005). It is clear that some degree of climate change during the next century is now inevitable (Houghton *et al.* 2001).

Hence, individuals, organisations and society as a whole will inevitably adapt to these changing conditions across a number of scales, sometimes successfully, sometimes unsuccessfully (Adger *et al.* 2005). Generally, a high reliance on natural resources, limited ability to adapt financially and institutionally, low per capita Gross Domestic Product (GDP) and high poverty are the key challenges for crop productivity in East Africa (Thomas & Twyman 2005). On the other hand, it shows that adaptation options for climate change in the Eastern region are largely drawn from the successful experience that community or system possesses in coping with past and/or ongoing climatic stresses (Dessai *et al.* 2005)

7.1 Climate Change Impacts on crop production in East Africa

Empirical studies have established that climate change has direct and indirect impacts on crop growth and development. For example, higher atmospheric concentrations of CO_2 have a direct impact on crops by altering photosynthesis and the efficiency of water use. In contrast, mean temperatures affect crop duration (Challinor *et al.* 2005c), whilst temperature extremes during flowering can reduce the grain or seed number (Wheeler *et al.* 2000; Challinor & Wheeler 2007). As a result, climate change will have distinctive implications on crop production in East Africa (Collier *et al.* 2008). Hence, while certain areas in Tanzania are projected to experience negative impacts, other areas are projected to experience positive impacts under climate change (Agrawala *et al.* 2003). Mwandosya *et al.* (1998) estimate that northern and south-eastern parts of Tanzania will likely experience an increase in rainfall ranging from between 5% and 45% under doubling of carbon dioxide, while central, western, south-western, southern parts, and eastern parts of the country and southern highlands might experience a decrease in rainfall of 10% to 15%.

It is well known that the sensitivity of agricultural systems to climate change differs between systems depending on whether they are temperature- or water-limited, and whether they are operating near their optimum or not (Gregory *et al.* 2009). Fuhrer (2006) concluded that there is ample evidence to demonstrate the sensitivity of agricultural systems to climate change and that the range of effects on potential productivity was from extremely negative in areas that were already water-limited to positive in areas that were temperature-limited. Photo 25 shows healthy produce that is threatened by climate variability.

Similarly, the effects of climate variability and change on crop production and agricultural systems are also location-specific and more importantly, societal-specific with countries and groups with low income and limited adaptive capacity being areas facing significant threats from climate change (Von Braun 2007). In particular, agricultural or crop failures in East Africa will be increased by climate change although the size of the effect is affected more by socio-economic factors than by climate change per se (Easterling *et al.* 2007). While most of the effects of climate change in East Africa are detrimental to crop production, some are potentially favourable. For instance, for maize, there are clear decreasing yield trends in the upper north-west of the region (northern Uganda, southern Sudan) and in lowland areas of Kenya and Tanzania. Large tracts of maize fields in Tanzania (Photo 26) have already been affected by climate variability Increasing maize yields are found in the highland areas of central Kenya and the Great Lakes Region (Thornton *et al.* 2009).

There are a few areas in the north-western Uganda, where weather patterns in the years subsequent to 2000 were projected to bring about some increase in maize yields, only for these increases then to be wiped out in subsequent years and yields to decrease. In addition, there are a few areas in central Kenya and in the region of Kilimanjaro where yields are projected to be stable or decrease initially, only to increase in subsequent decades (ibid). Other simulations show that maize yields will decrease, as a result of higher temperatures and where applicable, decreased rainfall. The average yield decrease over the entire country is expected to be 33% but simulations predicted a decrease as high as 84% in the central regions of Dodoma and Tabora. Yields in the north-eastern highlands may decrease by 22% and in the Lake Victoria region by 17%. The southern highland areas of Mbeya and Songea are estimated to have decreases of 10-15% (Agrawala *et al.* 2003).

These reductions are due mainly to increases in temperature that shorten the length of the growing season and to decreases in rainfall. The agriculture sector thus may have both negative and positive impacts that could partially offset each other. However, maize production in particular might require particular attention for adaptation and mainstreaming responses, given that it is a critical food crop (ibid). The situation for beans is more dynamic. Yield trends in some of the highland areas of Uganda and Kenya are projected to be positive and negative in other places, particularly the coastal areas (Thornton *et al.* 2009).

Photo 25: Agriculture of one sort or another provides a source of livelihoods to about 80% of Tanzania's population.

Likewise, it is projected that the carbon fertilization effect will be positive and East Africa appears likely to benefit from wetter climates (Warren *et al.* 2006). However, the most general implication is that the consequences will depend upon the capacity to adapt to change. Even the potentially favourable carbon fertilization effect will not benefit those farmers currently growing non-responsive crops unless they switch to responsive crops (Collier *et al.* 2008). The adverse effects, such as the increased incidence of drought, could have disastrous consequences unless appropriate response action is taken (ibid).

Currently, the effects of sudden shocks such as drought are felt, in addition to other ongoing, long-term stresses. The result observed is that the long-term stresses deplete household resilience so that the employment of coping and adapting strategies that might be available to other better-prepared communities to deal with sudden shocks, is at too high a cost or, simply, unavailable(Gregory *et al.* 2009). Also, a range of regional and global political and economic factors including high food prices, legacies of structural adjustment, government policies, conflict, and policies on genetically-modified foods, and poor responses to the HIV/AIDS pandemic(Vogel & Smith 2002) reduces the resilience of the communities to cope with the shock of drought (Gregory *et al.* 2009).

Studies suggest that while many crops may respond positively to increased atmospheric CO_2 concentrations (Long et al. 2004), the associated effects of higher temperatures and altered patterns of precipitation will reduce crop yields (Easterling et al. 2007). However, it is widely recognized that these projections are likely to represent an overestimate in actual field and farm level responses because they are derived from experiments and crop models that do not necessarily take limiting factors such as pests and pathogens, competition, nutrient competition and soil water fully into account (Gregory et al. 1999; Tubiello et al. 2007b)

Photo 26: Maize farming is an important activity in most rural areas of Tanzania.

Because of the fundamental effects of radiation, temperature and water on the growth of plants (Hay & Porter 2006), it is unsurprising that there has been considerable research to understand the effects of climate and climate change on crop production (Gregory *et al.* 2009). The likelihood of some regions benefiting from changed climate while others suffer has been highlighted by a number of modelling exercises that combine biophysically based crop models with global simulations of climate (ibid).

Overall, the results of this and subsequent work demonstrated that climate change would benefit the cereal production of developed countries more than the developing countries, even if cropping practices evolved to allow more than one rain-fed crop per year (Fischer *et al.* 2002, 2005). While crop biomass is predicted to increase in response to elevated CO_2 concentrations under many circumstances, it is also recognized that crops and soils may subsequently become nutrient limited, especially in terms of nitrogen availability (Diaz *et al.* 1993). The increased use of legumes within arable rotations has therefore been of considerable adaptation interest (Tubiello *et al.* 2007b) since legumes can increase N_2 fixation in response to elevated CO_2 (Soussana & Hartwig 1996).

Available literature also suggests that increased rainfall can have substantial effects on insect populations (Gregory *et al.* 2009). For example, Staley *et al.* (2007) investigated the impacts of enhanced summer rainfall and drought conditions on soil-dwelling *Agriotes lineatus* (wireworms) in grassland plots. Wireworms are damaging pests of crops such as potatoes, especially when planted on land taken out of grass (Johnson *et al.* 2008) and there is speculation that they are likely to become more of a problem as a result of climate change.

Studies conducted by Parker & Howard (2001) and Staley *et al.* (2007) established rapid increase in the population of wireworms in the upper soil as a consequence of enhanced summer rainfall events compared to ambient and drought conditions. In the seasonally arid regions of the developing world, people are particularly vulnerable to inter-annual and intra-seasonal rainfall variability, through dependence on rain-fed agriculture (Haile 2005; Cooper 2004). Hence, the potential benefits of climate forecasting may be particularly high in tropical regions, where there may be strong relationships between climate and impact variables such as crop yield (WCRP 2007)[18]

The impacts of pests and diseases on crop production in current conditions are well known, but the consequences of climate change on pests and diseases are complex and, as the preceding descriptions attest, are still only imperfectly understood (Gregory *et al.* 2009). Scherm *et al.* (2000) highlighted the importance of pests and diseases both as important yield-reducing components and as early indicators of environmental changes because of their short generation times, high reproductive rates and efficient dispersal mechanisms. Attempts have been made to model the effects of changing climate on the distribution of pests and pathogens, particularly using climatic mapping to

18 WCRP: World Climate Research Programme

delineate potential distributions based on the concept of the fundamental niche (Baker *et al.* 2000).

Some pests which are already present, but only occur in small areas or at low densities, may exploit the changing conditions by spreading more widely and reaching damaging population densities (Gregory *et al.* 2009). Aphids, for instance, key pests of agriculture, horticulture and forestry, are expected to be particularly responsive to climate change because of their low developmental threshold temperature, short generation time and considerable dispersal abilities (Sutherst *et al.* 2007). Together, the effects of changing climate and more variable weather suggest that pest and pathogen attacks are likely to be more unpredictable and the amplitude larger; and continue to be major constraints to food and agricultural production in parts of all regions of developing countries (FAO 2005; Gregory *et al.* 2009).

In spite of significant negative implication that climate change induced pests and diseases pose to crop production; control of plant pests still entails substantial use of pesticides, which have side effects on human health and the environment. This is particularly true for relatively poor farmers and farm labourers who cannot afford proper application equipment and appropriate personal protection (FAO 2005). The consequences for other elements of agro-ecosystems and crop yields are still uncertain and greater effort is required to integrate this science into estimates of actual crop productivity. The ability to include realistic impacts of pests and diseases in future climates has a direct link to considerations to crop productivity, crop yield and food security (Ingram *et al.* 2008). It is pointed out that more mechanistic inclusion of pests and disease effects on crops would lead to more realistic predictions of crop production on a regional scale and thereby assist in the development of more robust regional food security policies (ibid)

7.2 Crop Production Adaptation to Climate Change and Variability

7.2.1 An Overview

Empirical studies indicate that farmers have adjusted crop production methods and farming systems in response to past and present climatic changes, and many are now contemplating adapting to altered future climatic conditions. Some adaptation measures in Tanzania, Kenya and Uganda are undertaken by individuals, while other types of adaptation are planned and implemented by governments, sometimes in anticipation of change but mostly in response to experienced climatic events, especially extremes (Adger 2003; Kahn 2003; Klein & Smith 2003; FAO 2006; Ziervogel *et al.* 2008).

Much of the climate change adaptation literatures provide similar dichotomies in temporal and spatial scales in calibrating adaptation measures (Smith *et al.* 1996; Bohle 2001; Burton *et al.* 2003; Adger *et al.* 2005). However, much of available adaptations are perceived to be reactive, in the sense that they are triggered by past or current events, but some adapting measures are also anticipatory in the sense that they base on some assessment of conditions in the future (Adger *et al.* 2005).

Local-level adaptation actions are often portrayed as reactive, while higher-level institutions are assumed to plan in an anticipatory manner for adaptation through policies, programmes and, most recently through National Adaptation Plans of Action (Smith *et al.* 1996; Bohle 2001; Burton *et al.* 2003). These adaptation measures are geared to correspond to most negative impacts

Photo 27: Mixed farming - a strategy to maximise land use in the face of climate extremes..

of climate change on crop productivity in East Africa and Africa in general. Certainly, farmers have proved highly adaptable in the past to short- and long-term variations in climate and in their environment (see Photo 27). Key to the ability of farmers to adapt to climate variability and change is the access to relevant knowledge and information (Larsen & Gunnarsson-Ostling, 2009).

Arguably, adaptation to a changing climate is already occurring because of past and present observation, and future expectation, of natural and human-induced fluctuations of the climate (Dessai *et al.* 2007). Likewise, societies have a long record of adapting to the impacts of weather and climate through a range of practices that include crop diversification, irrigation, water management, disaster risk management, and insurance (Adger *et al.* 2007). For example, in most cases crops are planted during early rain season since the warmer temperatures in rain season provide the benefits of a longer growing season. The warmer temperatures at the end of the rains help the crops to ripen.

Common adaptation methods in agriculture adopted in East Africa include use of new crop varieties that are better suited to drier conditions (common crop varieties include cassava, sorghum and millet), irrigation, planting saline-tolerant crops, crop diversification, adoption of mixed crop and livestock farming systems, and changing planting dates (Bradshaw *et al.* 2004; Kurukulasuriya & Mendelsohn 2008; Nhemachena & Hassan 2007). In addition, farmers also switch between crops, alter the mix of crops they grow and change the timing of planting in the light of evidence they obtain of the growing season (O'Brien *et al.* 2000). For example, farmers may switch from maize to sorghum and/or cassava when there is a threat of drought or food insecurity and switch to rice or banana when rainfall is abundant. Crop switching can make a difference in productivity between years or seasons. Specifically, the existing adaptation activities in agricultural sector in Tanzania include the use of small-scale irrigation, the use of drought tolerant seed varieties, diversification of agriculture by growing different types of crops on different land units, and change crop rotation practices (URT 2007).

Other adaptation strategies include the practice of agricultural intensification (applying more inputs on units of land), agricultural extensification (bringing new units of land under cultivation), livelihood diversification (creating a portfolio of natural resource based and other livelihood activities) and migration (Scoones 1998: 9). Also strategies involve planting trees, soil conservation and use of different crop varieties, changing planting dates and irrigation. The use of different crop varieties is the most commonly used method, whereas use of irrigation is the adaptation least practiced among the major adaptation methods in East Africa.

Greater use of different crop varieties as an adaptation method is associated with the lower expense and ease of access by farmers, while the limited use of irrigation is attributed to the need for more capital and low potential for irrigation. These measures adopted by farmers in the region are likely to be practised in other regions as calibrated by other findings in the climate change adaptation literature (Bradshaw et al., 2004; Kurukulasuriya & Mendelsohn 2008; Maddison 2006; Nhemachena & Hassan 2007: 2008).

In the light of increased frequency and severity of droughts, change of staple crops to millet and cassava may be necessary in the inner part of the country. There is considerable uncertainty regarding the effects of climate change on the yields of most important cash crops such as coffee, cotton and tea. However, changing climate is unlikely to reduce their yields in the future to the same extent as that of maize (ibid.)

Existing measures aim most directly at climate adaptation with a focus on increasing agriculture's drought resistance. Essentially, drought-resistant crops are perceived in agricultural policies of East African governments as a principal means of addressing problems related to climate variability and drought in particular. Promotion of such species is integrated into national and district development policies, multi-sectoral policies, and sectoral policies in Kenya, Uganda and Tanzania (Ellis & Bahiigwa 2003; Eriksen, 2001). For example, Kenya focuses more explicitly on drought- resistant crops: a section in the NEAP (Kenya National Environmental Action Plan), regarding agriculture and food security, encourages famine and drought tolerant crops, in order to improve farmer resilience to drought. Also, the Kenyan Food Policy emphasizes preventing land degradation and encouraging use of drought-resistant crops in marginal areas that are vulnerable to climate variability (Eriksen 2001).

Irrigation is another priority adaptation measure implemented particularly in the agricultural sectors, in Kenya and Tanzania. Improvement and expansion of irrigation as well as water harvesting are identified as important measures to increase agricultural production especially in dry-lands. However, development of irrigation in areas with unreliable rainfall is a prominent district-level strategy in both Kitui and Same Districts.

However, efforts to enhance the drought resistance crop varieties of agriculture are in themselves unlikely to provide successful local adaptation to climate change. They face several constraints. Farmers are reluctant to adopt certain drought-resistant species, in part due to low market and consumption values, and in part due to high labour investment associated with cultivating these species. It is likely that successfully increasing cultivation of drought-resistant species requires numerous measures addressing social, economic and technical constraints (ibid).

These adaptations depend upon the decisions of individual firms and households (ibid). One reason for believing that much of the adaptation happens without government inducements is that the impact occurs only very gradually. However, private actors are presumed to respond appropriately to changing conditions depending on adequate information, appropriate incentives, and an economic environment conducive to investing in the required changes (ibid). Hence, the most promising strategy for Eastern Africa governments is to ensure that these three conditions are met (ibid).

The most information-intensive aspects of adaptation are likely to be changes between crops and crop varieties. Yet genetic modifications are useful means of speeding African agricultural adaptation through genetic modification. This has been put forward by Mendelsohn and Dinar (1999) that genetic adaptation strategies in place in some parts of Eastern Africa region include combinations of new hybrids, changes in sowing dates, and double cropping, using short-cycle maize varieties as a second crop along with lentils and a vetch-forage barley mixture. These strategies not only reduce the impact of increased temperatures on yields but also permit more intensive use of water and land (ibid). For example, Kaiser et al. (1993, a, b) consider adaptation practices such as crop mix, crop varieties, sowing and harvesting dates, and water saving technologies (shallow tillage). In terms of adaptation options, this simple characterisation suggests the need for more drought-tolerant maize varieties, coupled with management practices that can make use of the most available rainfall (such as water harvesting, for example). For bean production, the results suggest that a shift in bean cropping to higher elevations may be appropriate.

A particular limitation of adaptation studies in the agricultural sector stems from the diversity of climate change impacts and adaptation options but also from the complexity of the adaptation process. Many studies make the unrealistic assumption of perfect adaptation by individual farmers. In contrast, even if agricultural regions can adapt fully through technologies and management practices, there are likely to be costs of adaptation in the process of adjusting to a new climate regime (Dixon et al. 2003; Adger et al. 2007).

Hence, there is no doubt that adaptive capacity to climate change in East Africa, among individuals, institutions and governments is uneven across and within the region (Gregory et al. 2005; Adger et al. 2007). There are individuals and groups within all societies that have insufficient capacity to adapt to climate change (Dixon et al. 2003; Adger et al. 2007).

Noticeably, the capacity of smallholder farmer households in Kenya and Tanzania to cope with climate stresses is often influenced by the ability of a household member to specialise in one activity or in a limited number of intensive cash-yielding activities (Eriksen et al. 2005). In addition, women in subsistence farming communities are disproportionately burdened with the costs of recovery and coping with drought in Tanzania, Kenya and Uganda (See for example, Adger *et al.* 2007). Likewise, there are evidences that the determinants of adaptive capacity of smallholder farmers in Kenya and Tanzania are multiple and inter-related (Eriksen *et al.* 2005; Adger *et al.* 2007).

It is also established that the capacity to adapt is dynamic and influenced by economic and natural resources, social networks, entitlements, institutions and governance, human resources, and technology. There are also significant knowledge gaps for adaptation as well as impediments to flows of knowledge and information relevant for adaptation decisions. New planning processes are attempting to overcome these barriers at local, regional and national levels through the established National Adaptation Programmes of Action (NAPA) which specify adaptation strategies for the rapid changing climate in the eastern region and Africa in general (Adger *et al.* 2007).

7.2.2 Experiences from Tanzania

Field observations of crop production adaptation to climate change were done in two villages of Dodoma District (Laikala, Kongwa district and Chibelela, Bahi district). Findings illustrate some of the complexities of analysing climate-

Photo 28: Drought badly affects human socio-economic development.

change impacts in areas characterised by climatic variability where farmers are already coping with uncertainty coupled with ongoing trends (for example, endemic poverty, rapid population growth, etc.), which limit people's capacity to adapt (Thomas & Twyman 2005; Morton 2007).

There is a high degree of consensus in local observations of climate change across different social groupings. Shortened and highly unpredictable rainfall season; extreme winds, which now blow all year round due to loss of vegetation cover; stronger winds forcing rain clouds across the sky without letting them rain; more intense sunshine and heat. Increased drought as a result of less predictable and more intense sunshine and heat signify climate change in the area. Photo 28 illustrates the impoverishment of rural communities caused by the effects of drought. Also, communities identified a much colder period in June and July than in the past. These local observations about climate change are consistent with scientific projections (Mwandosya *et al.* 1998; Hulme *et al.* 2001; IPCC 2001) which suggest that Tanzania will warm by between 28°C and 48°C by 2100. Changes in temperature and rainfall are likely to prolong dry seasons and to worsen periodic droughts, particularly inland (Nelson & Stathers 2009).

The increasing unpredictability of the rainfall season has led to more people having to use oxen-drawn ploughs. Ploughing land using oxen is much faster than by hand and this speed allows for maximum use of the shortened, often intermittent rainy period for crop production (ibid). However, the poorest households can rarely afford to plough using oxen, and the wealthier owners prepare their own fields first. There are frequent crop failures and yield variability due to unpredictable rainfall, declining soil fertility and increased incidence of some pest and disease problems. Likewise, farmers have to replant bulrush millet and groundnuts more often, as rains are unpredictable, coming and then stopping abruptly, meaning that time and seeds are wasted and the quality of the crops affected (Nelson & Stathers 2009).

There are changes in the crops grown and the need to replant more frequently. Farmers have changed the balance of crops grown, with greater cultivation of drought-tolerant crops. They are choosing different faster-maturing sorghum varieties, because the rainy season is now so short that their traditional varieties cannot mature in time (ibid). Sesame and sunflower have been introduced following market demand and government advice, because they are more drought-tolerant. Cassava production has increased, because it is a drought tolerant food crop. Farmers are changing the balance of crops grown due to climate, but also government advice and market signals. Also, the communities claimed that some crops (maize, groundnuts, Bambara nuts, and cowpeas) are being grown more widely to earn cash and because women received training on improving storage (ibid).

7.2.3 Experiences from Uganda

Farmers in Uganda have established mechanisms to adapt to climate change through weather forecast information. This is happening because rainfall patterns differ from the north to the south due to climate variability and change, so do dominant crops and cropping systems. In the north, where rainfall is almost uni-modal, annual crops such as millet and other grains are prominent. Late-maturing varieties of sorghum, sesame and pulses are also planted, as are cassava, legumes and other vegetables (Phillips & McIntyre 2000).

Grains are usually planted first in March and then again in August or September, so that physiological maturity will occur in the dry season. Perennial crops, such as banana and coffee, dominate most of the cropping systems of the bimodal rainfall zone, although locally distinct cropping systems (e.g. grain in south-western and eastern highlands) do exist. Early-maturing grain crops (e.g. 120 days to maize dry harvest) and pulses are grown in both rainy seasons. Other commonly planted crops include sweet potato, potato, cassava and legumes (ibid).

Planting dates for annuals vary with the onset of the rains. Depending on soil moisture, maize may be planted from mid-August to mid-September in the first season and mid-February to mid-March in the second season. Beans are often planted well into April and October. Given the dependence of planting time and crop choice on rainfall distribution, there could be potential for utilizing forecasts of season arrival date and duration in crop management (ibid).

7.2.4 Challenges facing crop adaptation measures

A major challenge facing the agricultural sector in East Africa is how to establish crop adaptation strategies to climate change and variability, given the pace of a number of areas severely prone to drought. For example, it is very striking that in India 40% of the arable land is irrigated whereas in Africa the figure is just 4% (Stern 2006), and estimates for irrigated cropland in Tanzania as of 1999 was 3.3% (World Bank 2002) . This lack of irrigation, together with the rest of infrastructure, is one of the reasons why Africa is so vulnerable to climate change (Stern 2006).

Under climate change, inherent uncertainties in the predictability of climate limit the precision with which impacts can be assessed (Challinor *et al.* 2009). Furthermore, the response of crops to elevated carbon dioxide is not known with precision at field and larger scales (Ewert *et al.* 2002; Tubiello & Ewert

2002). Quantification of uncertainty is therefore an important endeavour in climate impacts research (Challinor *et al.* 2009b). Estimates of ranges of yield impacts vary across studies (the review of Luo & Lin 1999).

The simulated responses of maize in Africa to a doubling of carbon dioxide, for example, can be as broad –98% to +16%, or as narrow as –14% to –12%. These ranges have been determined using different methods and are therefore not directly comparable (Challinor *et al.* 2007). The response of crops to any projected climate also contains uncertainties (Mearns 2003). Inputs to crop models, such as the choice of variety and planting date, can be varied in order to produce an ensemble of crop simulations (Jagtap & Jones 2002; Irmak *et al.* 2005).

Climate change adaptation planning has been hindered by the unavailability and unreliability of climate predictions, which would allow the formal estimation of the probability of impacts, resembling weather forecasts and famine early warning systems (Downing 2003). Also, although the breeding of new cultivars with improved yields under future climate is a potentially crucial adaptation option, the basis on which any new cultivars are developed will depend on the nature and extent of climate change in any specific region or cropping system (Wilby *et al.* 2009).

Ingram *et al.* (2008) outline three major challenges for agronomic research in the climate or food security debate that are relevant in Africa under the changing climate. These include, the understanding better on how climate change will affect cropping systems (as opposed to crop productivity); to assess technical and policy options for reducing the deleterious impacts of climate change on cropping systems while minimizing further environmental degradation; and to understand how best to address the information needs of policy-makers and report and communicate agronomic research results in a manner that will assist the development of food systems adapted to climate change. According to Gregory *et al.* (2009), agricultural research community should think more about the issues of scale and how to translate findings at plot-scale over a few seasons to larger spatial and temporal scales and to the issues of food security (ibid).

CONCLUSION

There are many complex processes and interactions that determine crop yield under changing climate. These include the response of crops to mean temperature, the interaction between water stress and CO_2, and the interaction between ozone and a range of environmental variables. As a result of this and of the importance of scale and geography in determining crop productivity, perhaps the greatest challenge for future syntheses of knowledge on the response of crops to climate change is the balance between generality and specificity in region and scale (Challinor *et al.* 2009b).

The implication for adaptation therefore may be to not only cushion adverse impacts, but also to harness positive opportunities. This suggests consideration of an enhanced portfolio of linked-adaptation responses – for example a strategic shift from maize to cash crops over the medium term, and inter-basin transfers in the case of water resources. Such strategic shifts however may entail economic and dislocation costs – and therefore require careful screening, particularly with regard to their effects on equity and rural livelihoods. More rigorous testing of particular crop and stream-flow projections may also be advisable prior to undertaking such adaptation responses (Agrawala *et al.* 2003).

While many studies have demonstrated the sensitivities of plants and of crop yield to a changing climate, a major challenge for the agricultural research community is to relate these findings to the broader societal concern with food security (Gregory *et al.* 2009). Evidence for the impact of climate change on crops and adaptation strategies in east Africa are being documented. However, the documentation of adaptation strategies in east Africa are not well elaborated and poorly understood in the context of climate change. Hence, adequate studies on climate change effects in crop adaptation strategies would lead to more realistic predictions of crop production on a regional scale and thereby assist in the development of more robust regional food security policies (ibid).

The performance of agriculture in the past has been very weak: output per hectare over the last 40 years is virtually unchanged whilst population has trebled (Stern 2006). For viable adaptation, stronger growth in agriculture in Africa requires action across a number of fronts. These fronts include infrastructure, particularly transport and irrigation, agricultural extension and the development of improved varieties of crops and poverty rights. These rights need to give incentives to invest in the land and farm, including for women, access to micro-credit, physical security so that long-term rewards to investment can be realised better forecasting and response to weather and pests.

Assessments of the relationships between crop productivity and climate change rely upon a combination of modelling and measurement. In contrast, the challenges associated with climate change impacts and farmers adaptation are still unaddressed. It is argued that the generation of knowledge for policy and adaptation should be based not only on syntheses of published studies but also on a more synergistic and holistic research framework that includes, among other things:

(i) reliable quantification of uncertainty;

(ii) techniques for combining diverse modelling approaches and observations that focus on fundamental processes; and

(iii) judicious choice and calibration of models, including simulation at appropriate levels of complexity that accounts for the principal drivers of crop productivity, which may well include both biophysical and socio-economic factors. It is argued that such a framework will lead to reliable methods for linking simulation to real-world adaptation options, thus making practical use of the huge global effort to understand and predict climate change (Challinor *et al.* 2009).

When considering adaptation, it is important to consider how weather and crop yield forecasts will be used and what spatial and temporal scales will be the most appropriate for the users. Useful weather/climate forecasts can range from a few days ahead for some crop management decisions to decades in the future for infrastructure and strategic planning (Challinor et al. 2009). For example, ensemble climate modelling can be used with crop models in order to predict crop yield a season ahead of the harvest (Challinor et al. 2005c). Information should also be provided in relevant formats (Stone & Meinke 2005).

Whether the information best suited to users is based on computer-intensive systems or on less high-tech systems such as observational networks and capacity building depends to a large extent on the particular users considered (Patt *et al.* 2009). In Africa, for example, a prudent way to address the threat of climate change may be to focus on strategies for coping with climate variability, rather than longer term climate change (Washington *et al.* 2006). This may mean a greater focus on in situ and remotely sensed observations as well as consideration of the multiple stresses that act on food security (Gregory *et al.* 2005; Haile 2005; Verdin *et al.* 2005).

PART
3

CHALLENGES AND OPPORTUNITIES IN ADDRESSING: CLIMATE CHANGE IMPACTS

FACTORS INFLUENCING CHOICE OF RESPONSE STRATEGIES TO CLIMATE CHANGE AND VARIABILITY IN ZIMBABWE AND ZAMBIA

8.0 Introduction

It is important to understand factors that influence farmers' adaptation, as this is expected to point towards understanding the opportunities and challenges that exist in adapting to climate change and variability. There are a host of factors that influence the adaptive capacity of farmers in the face of a wide range of hazards. These factors have been classified by scholars into two categories, namely, generic and specific. While education, income and health have been considered to be generic, specific determinants to particular impacts such as droughts and floods may relate to institutions, knowledge and technology (Brooks *et al.* 2005; Downing 2003; Tol & Yohe 2007; Yohe & Tol 2002).

Numerous studies have pinpointed that institutions and their effective functioning play a critical role in successful adaptation (Adger *et al.* 2007; Reid & Vogel 2006). It is therefore important to understand the design and functioning of such institutions which include both formal and informal institutions. For instance, Reid & Vogel (2006) established that these institutions may be conceptualized as farmers and local community groups, public and government institutions and local organizations. Therefore, links to these institutions also shape the adaptive capacity of a household. Also related to these institutions are the enhanced communication of climate-related information and the development of social networks which can assist farmers to know times of higher probability of success in adaptation (NOAA 1999; Stern & Easterling 1999). There is need to strengthen institutions and facilities that disseminate weather information for the benefit of farmers. However, there is a challenge in this respect as there is evidence that seasonal forecasting in Africa has not reached optimum levels (Thiaw *et al.* 1999).

Other factors influencing adaptive capacity include demographic factors, dependence on agriculture and natural ecosystems and resources, poverty and inequality (Adger *et al.* 2003; Reid & Vogel 2006). For example, low adaptive capacity of Africa is due in large part to the extreme poverty of many Africans. Risk spreading is accomplished through kinship networks, pooled community funds, insurance and disaster relief. In many cases the capacity to adapt is increased through public sector assistance such as extension services, education, community development projects and access to subsidized credit. This further underscores the role of institutions and social networks. Societies have inherent capacities to adapt to climate change. These capacities are bound up in the ability of societies to act collectively, thereby emphasizing the importance of social networks for small scale farmers. The adaptive capacity of societies depends on the ability to act collectively in the face of the threats posed by climate variability and change (Adger *et al.* 2003).

In light of the above background, this chapter presents factors that influence adaptation for small holder farmers in Zimbabwe[19] and Zambia.

8.1 Data collection and analysis

Data analysed for this chapter were collected in a questionnaire survey across 720 households in two districts of Zimbabwe and Zambia and entered into the Statistical Package for the Social Sciences (SPSS). Data collected include demographic factors, response strategies to climate change and variability and vulnerability context, among others. The logistic regression was then used to analyze factors that influence whether a household adapts using a certain technology or not. The data collected through the questionnaire survey were complemented by data collected through focus group discussions.

The complex factors that determine the success of response strategies have already been identified as including access to resources, household size and composition, access to resources of extended families and the ability of the community to provide support, among others (De Waal 2005; Mutangadura, Mukurazita & Jackson 1999). Based on existing literature, five groups of factors were tested (see Table 3), which are: (i) demographic factors; (ii) access to information and technologies; (iii) assets and resources; (iv) membership of groups is used as a proxy for access to informal institutions; and (v) vulnerability; whereby it is hypothesized that farmers in Zimbabwe are more vulnerable due to the economic climate in that country. A logistic regression is then used to analyze factors that influence the use of different coping and adaptation strategies. Therefore, the model is measuring factors that affect use of various coping and adaptation strategies.

The dependent variable, which is Y, is either an adaptation or coping strategy presented in Table 4 or the conservation farming methods presented in Table 5. The general model is:

---

19 This chapter is based on the same research project that was highlighted in Chapter Six. The study was conducted in Lupane and Lower Gweru districts (Zimbabwe) and Monze and Sinazongwe (Zambia).

$$Y = b0 + bX1 + bX2 + \ldots\ldots + bXn$$

Y=either 0 or 1 where 0 means no use of strategy and 1 represents use of strategy.

Selected strategies from adoption of conservation methods and other strategies and how they are influenced by different factors are presented in Tables 4 and 5. The most dominant strategies in the two countries are selected for the model (see Section 8.2).

8.2 Response strategies in the study areas

8.2.1 Conservation farming methods used

Use of conservation farming methods was found to be dominant as a response to both floods and droughts, follow up questions were asked to establish the specific methods that farmers use (Figure 2). All the methods, with the exception of winter ploughing and intercropping are mostly practiced by households in Monze district (Zambia). Winter ploughing is most common in Lower Gweru and intercropping in Lupane (both in Zimbabwe). The growing of drought tolerant crops and varieties is the most dominant in all the districts, which is congruent with farmers' perceptions that precipitation has declined over the years. This also supports the notion that farmers respond to climate variability based on their perceptions of this variability.

In Sinazongwe (Zambia), farmers plant drought tolerant crops such as millet, cotton, sorghum and cow peas, which they plant late in the season around January, in order to take advantage of the late rains that they now expect for each season. In Monze, they also plant late maturing varieties[20]. In Lupane, they grow sorghum, millet and gourd-like vegetables (*mashamba* and *amajodo*). Use of crop residue is least common in Lower Gweru as these farmers indicated that in a bad season, farmers keep crop residue for stock feed and not for their fields.

8.2.2 Other response strategies

Farmers highlighted that there is food insecurity that largely emanates from climate variability. It was therefore important to establish how they respond to this food insecurity. It is illustrated in Figures 3 and 4 that livestock sales, renting out land and the selling of firewood are the only strategies that are practiced more in Monze and Sinazongwe than in Lupane and Lower Gweru. The rest of the strategies highlighted in these figures are most dominant in

---

20 Examples of the late maturing varieties used include MRI, 624, Seed Co. 701 and 709

Zimbabwe districts. This evidence suggests that farmers in Zimbabwe districts engage more in livelihood diversification than those in Zambia districts. This fact is not unexpected considering evidence from previous sections that Zimbabwe districts are harder hit with multiple stressors than Zambia districts. This contradicts the notion that vulnerable populations have little recourse in the face of food insecurity.

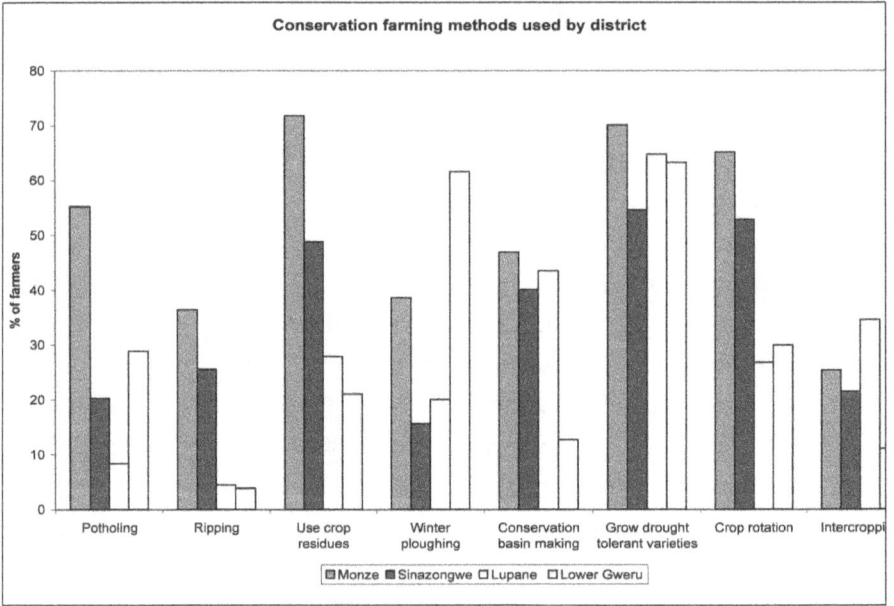

Figure 2[21]: Conservation farming methods used by farmers in the study districts in Zimbabwe and Zambia

---

21 Data presented in the figures in this section were collected between March 2008 and March 2009

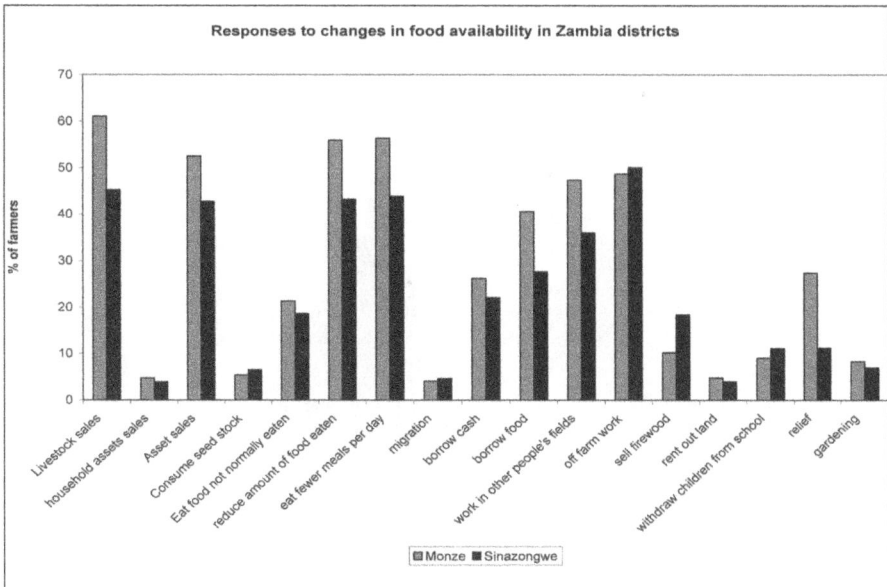

Figure 3: Responses to changes in food availability in Zambia by district

In addition, adaptations such as diversifying livelihoods are not responses unique to climate disturbances, and all are embedded in the full range of livelihood changing factors (Thomas *et al.* 2007). Furthermore, households facing regular episodes of food insecurity have been found to develop complex strategies for dealing with these events (Liwenga 2003; Reardon, Malton & Delgado 1988; von Braun, Teklu & Webb 1998). The implication is that their livelihoods are more vulnerable to production shocks than less vulnerable households and that there are multiple coping strategies available to them (Phillips 2007; Swift & Hamilton 2001). Diversity, therefore, could have been used as an insurance mechanism in an unpredictable environment or can be a necessity in the face of immediate food insecurity (Hulme 2001).

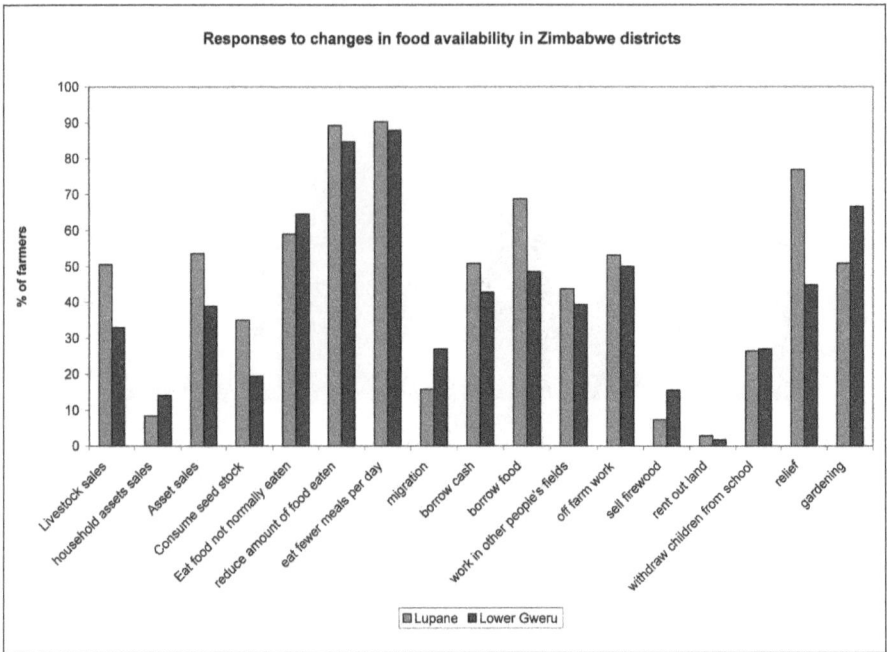

Figure 4: Responses to changes in food availability in Zimbabwe by district

8.3 Influencing factors in Zimbabwe and Zambia

Age of household heads can influence the livelihood strategies pursued by the households. According to Ellis *et al.* (1998) and Bebbington (1999) family life cycle characteristics such as age, education, and the number of family members can influence household's and individual's objectives, such as risk management practices, consumer preferences, and/or strategies available to cope with shocks. In the model in Table 5, there is a positive and significant correlation between age of household heads and use of potholing. Similarly, in studies done in Ethiopia and Kenya, older household heads were more likely to adopt climate change adaptation measures in the form of conservation farming technologies than were younger ones (Yesuf *et al.* 2008; Anjichi 2007). In addition, Table 4 shows that younger households are more likely to eat food that they normally would not eat during times of food shortage as well as resort to gardening. Older households, on the other hand, are more likely to reduce the amount of food they eat when faced with food shortages resulting from climate change and climate variability.

This could be the case because reports from group discussions revealed that in drought situations, younger men and women migrate to other areas and leave their homes to engage in gold panning and other trading activities such as cross border trade, particularly in Zimbabwe.

The same reports indicated that younger farmers have less land at their disposal than older farmers, which could explain why younger farmers engage more in gardening, an activity that requires less space, either close to the homesteads or in the wetlands.

Table 3: Definition of variables influencing adaptation

Demographic	
X1- Age of household head	Age of household head in years
X2- Sex of household head	Sex of household head 0=Male, 1=Female
X3 -Marital status	Marital status of household head 0=Unmarried, 1=Married
Access to information and technologies	
X4- Access to weather information	Whether household is accessing weather information 0=No, 1=Yes
X5 -Participation in training	Whether any member of the household has participated in any agricultural or climate related in training 0=No, 1=Yes
X6 -Education level of household head	Education level of household head 0=No formal education, 1= Have primary, secondary or tertiary education
X7 –Perception of climate variability	Observations of any weather changes by household in the past years 0=No, 1=Yes
Assets and Resources	
X8 -Land owned in previous season	Size of land owned by household in previous season in ha
X9 -Cattle numbers	Number of cattle owned by household

X10- Poverty	Household own perception of their level of poverty 0=Poor, 1= Medium or Rich
Institutions	
X11- Membership of group	Whether any member of the household belongs to a farmer group/association 0=No, 1=Yes
Vulnerability	
X12- Harvest duration in good year	Duration of harvest of main cereal in a good season in months
X13- Harvest duration in bad year	Duration of harvest of main cereal in a bad season in months
X14 -Country	0=Zimbabwe, 1=Zambia

Results (Table 4) further show that female headed households are more likely to borrow food and cash than male headed households. This is consistent with the hypothesis that female headed households are more likely to engage in erosive strategies than male headed households.

Table 4: Factors influencing responses to climate variability and its outcomes

Variable	Asset sales	Food not normal-ly eaten	Reduced food quanti-ties	Borrow food/cash	Off farm work	Garden-ing
Demographic						
Age of house-hold head	0.004 (0.006)	-0.013** (0.006)	0.014** (0.007)	0.011 (0.011)	-0.003 (0.006)	-0.013* (0.007)
Sex of house-hold head	-0.293 (0.228)	0.032 (0.261)	-0.307 (0.280)	1.224*** (0.387)	0.099 (0.232)	0.121 (0.295)
Marital status	-0.204 (0.215)	-0.170 (0.248)	-0.182 (0.255)	-0.399 (0.368)	-0.085 (0.220)	0.333 (0.286)
Access to information and technology						
Access to weath-er information	0.081 (0.188)	-0.645*** (0.205)	-0.013 (0.245)	-0.781** (0.368)	-0.080 (0.193)	-0.150 (0.227)
Participation in training	0.047 (0.178)	0.211 (0.206)	0.204 (0.223)	0.437 (0.347)	0.307* (0.183)	0.111 (0.224)

Education of household head	-0.050 (0.236)	-0.313 (0.267)	0.288 (0.297)	0.632 (0.480)	0.422* (0.241)	0.006 (0.292)
Perception of climate variability	0.010 (0.241)	-0.124 (0.275)	-0.566* (0.330)	-0.148 (0.480)	-0.292 (0.249)	0.389 (0.289)
Assets and resources						
Land owned in previous season	-0.082** (0.041)	-0.019 (0.050)	-0.039 (0.044)	-0.105 (0.078)	-0.036 (0.041)	0.051 (0.057)
Land cultivated in previous season	0.129** (0.052)	0.126** (0.063)	-0.060 (0.065)	0.070 (0.116)	-0.060 (0.053)	-0.177*** (0.067)
Cattle numbers	0.020 (0.018)	-0.031 (0.020)	-0.027 (0.021)	-0.075* (0.044)	-0.090*** (0.022)	0.040* (0.022)
Poverty	0.602*** (0.181)	-0.092 (0.210)	0.057 (0.229)	0.573 (0.361)	-0.250 (0.186)	0.048 (0.229)
Institutions						
Membership of group	0.096 (0.179)	0.147 (0.206)	0.232 (0.227)	0.653* (0.346)	-0.215 (0.183)	0.272 (0.219)
Vulnerability						
Harvest duration in bad year	-0.024 (0.033)	-0.162*** (0.040)	-0.127*** (0.038)	-0.065 (0.058)	-0.079** (0.034)	0.003 (0.044)
Country	0.153 (0.205)	-1.721*** (0.229)	-2.238*** (0.270)	20.859 (0.734)	0.036 (0.210)	-2.892*** (0.288)
Constant	-0.614 (0.512)	2.231*** (0.595)	2.626*** (0.639)	-22.834 (0.734)	0.958* (0.529)	0.409 (0.652)

*, **, *** significant at the 10%, 5%, 1% levels

This coping strategy is considered to be a 'dangerous' one as the households concerned will have to return the food or cash soon after harvests, leaving them more vulnerable as they have less food or cash to last them the season and to be prepared if disaster strikes (Young & Jaspars 1995).

Table 5: Factors influencing use of conservation farming methods in climate variability and its outcomes

Variable	Pothol-ing	Use of crop residues	Wint. PL	Con. Bas. making	Grow-ing D.T crops	Changing crops
Demographic						
Age of house-hold head	0.015** (0.007)	-0.007 (0.007)	0.005 (0.008)	-0.002 (0.008)	-0.016 (0.008)	-0.008 (0.006)
Sex of house-hold head	0.455* (0.254)	-0.231 (0.259)	0.447* (0.249)	0.148 (0.246)	-0.109 (0.242)	0.593** (0.253)
Marital status	0.234 (0.240)	-0.081 (0.242)	0.530** (0.240)	-0.013 (0.228)	-0.161 (0.229)	0.209 (0.233)
Farming experi-ence of house-hold head	-0.002 (0.003)	0.001 (0.002)	-0.004 (0.008)	0.007 (0.009)	0.008 (0.008)	0.001 (0.002)
Access to information and technology						
Access to weath-er information	0.050 (0.075)	-0.020 (0.066)	0.124 (0.205)	-0.018 (0.039)	0.357* (0.202)	0.064 (0.083)
Participation in training	0.350* (0.208)	0.598*** (0.201)	0.271 (0.199)	0.946*** (0.199)	0.295 (0.194)	0.327* (0.195)
Education level of household head	0.032 (0.280)	-0.806*** (0.274)	0.165 (0.272)	-0.313 (0.258)	0.117 (0.254)	-0.162 (0.268)
Perception n of climate vari-ability	0.247 (0.296)	0.900*** (0.297)	0.184 (0.272)	-0.135 (0.261)	0.251 (0.257)	0.960*** (0.296)
Assets and resources						
Land owned in previous season	0.086* (0.045)	0.008 (0.046)	0.089** (0.043)	0.050 (0.044)	0.033 (0.046)	0.098** (0.049)
Land cultivated in previous season	-0.210*** (0.074)	-0.025 (0.064)	-0.123** (0.060)	-0.043 (0.060)	-0.018 (0.059)	-0.026 (0.064)
Cattle numbers	-0.057** (0.025)	0.018 (0.019)	0.024 (0.018)	-0.002 (0.019)	0.044** (0.022)	0.013 (0.019)
Poverty	-0.147 (0.210)	0.551*** (0.207)	0.283 (0.197)	0.137 (0.198)	0.349* (0.198)	0.404** (0.200)
Institutions						

Membership of group	0.033 (0.208)	-0.493** (0.204)	0.569*** (0.190)	-0.483** (0.197)	0.034 (0.197)	0.160 (0.199)
Vulnerability						
Harvest duration in good year	0.026 (0.028)	-0.065** (0.027)	-0.007 (0.025)	0.001 (0.026)	-0.027 (0.026)	-0.029 (0.026)
Harvest duration in bad year	0.100** (0.044)	0.127*** (0.044)	0.051 (0.041)	0.006 (0.042)	0.101** (0.045)	0.092** (0.042)
Country	0.781*** (0.233)	1.275*** (0.218)	-0.746*** (0.227)	0.581*** (0.216)	-0.543** (0.228)	1.105*** (0.217)
Constant	-2.924*** (0.635)	-0.863 (0.594)	-1.983*** (0.586)	-1.056* (0.569)	0.504 (0.566)	-2.243*** (0.595)

*, **, *** *significant at the 10%, 5%, 1% levels*

This may leave households in a cycle of poverty from one season to the next. Literature shows that this finding has to do with unequal access to resources by females in most African countries. Females have been found to have less access to resources such as land, property and public services (Agarwal 1991; Nemarundwe 2003; Njuki *et al.* 2008; Thomas-Slayter *et al.* 1995).

It is therefore an unexpected result in the analysis that there is a positive and significant relationship between sex and adaptation strategies such as potholing, winter ploughing and changing crops (Table 5). This implies that households that are headed by females are more likely to engage in these adaptation strategies than those headed by males. This is inconsistent with other studies that have highlighted that households headed by males are more likely to adopt technologies such as putting up soil erosion structures, fallowing and use of fertilizer and manure in Kenya, Cote D'Ivore and Burkina Faso (Adesina 1996; Matlon 1994; Njuki *et al.* 2008). Be that as it may, this study also finds that females (41%) are significantly more engaged in group activities than males (35%).

It would therefore appear that group membership influences farmers' adaptation, implying that this domination of farmers' groups by females may lead to subsequent adaptation by female (49%) who are also significantly involved in formal and informal training activities. What is emerging here is that social networks and relations may offset the vulnerability of farmers. Group membership is considered to be one of the elements of social capital (Njuki *et al.* 2008). There was a similar assumption therefore that members of farming groups and associations should be in a position to employ adaptation strategies if they can adopt technologies (Dube *et al.* 2007). Consistent therefore, is the result that there is a positive and significant relationship between group membership and winter ploughing. It may therefore be worrying that the social fabric, which is important, has been found to be deteriorating due to economic realities.

Indicators of access to information and technologies as expected to influence the use of different strategies. Households that had heads of households with primary, secondary or tertiary education were more likely to engage in off-farm work as were those who had training in agriculture. Higher education of course increases the likelihood of people having opportunities for off farm employment and such households are therefore more pre-disposed to using income from these sources to cope with food unavailability. Similarly, it is not surprising that farmers with access to climate information were less likely to use erosive strategies such as borrowing food or cash or eating food that is normally not eaten including eating seed. Srivastava & Jaffe (1992) argue that access to weather information is critical for the planning of farmers' agricultural activities and enhancement of their adaptive capacity.

Therefore, the negative relationship between education and use of crop residues may be puzzling at face value. By implication, farmers who are educated are less likely to use crop residues. This is unexpected because the level of education of household heads has been documented to be an important determinant of adoption of technologies and an educated farmer can readily access relevant information (Anjichi *et al.* 2007; Asfaw *et al.* 2004; Jayne *et al.* 2004; Mapila *et al.* 2002; Nkamleu 2007; Yesuf *et al.* 2008;). However, it is important to understand that traditionally, residues are used to feed livestock and with no information on potential benefits of retaining residues, farmers are likely to continue prioritizing livestock feeding. Understandably, reports from focus group discussions show that pastures have been affected by the recurrent droughts, possibly explaining why farmers may not be too keen to use crop residue in their fields.

Similarly, participation in training sessions by farmers has a positive and significant influence on their choice of adaptation strategies such as potholing, crop residue use and changing crops. This is consistent with literature that underscores the role played by formal and informal institutions in addressing the issue of climate change adaptation by farmers (Yesuf *et al.* 2008). Similarly, survey results indicate that most of the organized training that has taken place in Zambia and Zimbabwe has been largely on conservation farming technologies (40%) and crop management (21%). In addition, participants in focus group discussions reported that government extension agents (AGRITEX[22] in Zimbabwe and MACO[23] in Zambia) assist them with extension services that address climate change adaptation in conjunction with seed houses that breed seed varieties suitable for varying areas and climates. A study by Deressa *et al.* (2008) in the Nile Basin similarly posits that access to formal agricultural extension, farmer to farmer extension and access to weather information guarantees that farmers apply adaptation measures on their farm in comparison to those that do not have this access. Similarly, access to weather information is positively related to the growing of drought tolerant crops and varieties. This is understandable as farmers would grow drought tolerant crops when they have been alerted of a drought or inconsistent rains. It is therefore a disturbing fact that government services have been found to be on the decline given the critical role that they play in enhancing adaptation.

There is a positive and significant relationship between the size of land cultivated by a household and eating food not normally eaten. This suggests that the more land farmers cultivate, the more they eat food they do not normally eat. Where farmers with larger pieces of land are expected to adopt technologies more (Njuki *et al.* 2008) and by implication cope less, it has been documented in a study done in Zambia and other Southern African countries such as Malawi and Mozambique that farmers cultivating more land are likely to use less amounts of fertilizer across the large area vis intensifying and targeting use within a small area (Njuki & Mapila 2007). By implication, productivity may be low in larger pieces of land, explaining why farmers who cultivate smaller pieces of land are less likely to engage in erosive strategies. Moreover, the negative relationship between land cultivated and gardening implies that farmers cultivating less land are more engaged in gardening, possibly to supplement food stocks.

There is a negative and significant relationship between land cultivated in the previous season and potholing and winter ploughing. The suggestion herein is that as farmers increase land under cultivation, they are less likely to adapt by engaging in winter ploughing and potholing. Engaging in potholing and winter ploughing entails extra labour for farmers as they have to increase the number of times that they plough. Therefore, having more land under cultivation would require draught power and extra effort from farmers. Moreover, report from group discussions highlighted that farmers have reduced land under cultivation for other various reasons that include high

22 Agricultural, Technical and Extension Services
23 Ministry of Agriculture and Co-operatives

input price, climate variability and inadequate access to draught power, among others. A similar trend reported by C-SAFE[24] showed that more than forty per cent of Zimbabwean rural households in 2003 were not cultivating as much land as they previously had (Senefeld & Polsky 2005).

Land owned has a positive and significant influence on use of potholing and changing crops as adaptation strategies. This result suggests that the size of the farm influences adoption of technologies as the larger the land, the higher the chances and space for engaging in changing crops and potholing. Farmers with larger pieces of land are more likely to experiment and to have a broader crop mix (Njuki *et al.* 2008).

The period harvests from the previous season lasts is a determinant of whether farmers adapt to climate variability or not. Results from the model show that harvests in drought years positively and significantly influence the employment of strategies such as changing crops, growing drought tolerant crops, use of crop residues and potholing. This suggests that when there has been a bad cropping season, farmers whose harvest lasts longer engage more in adaptation strategies than those whose harvest lasts for shorter periods of time. It is generally assumed that farmers who are somewhat food secure tend to be more resource endowed, older and more labour secure than those who are less food secure, explaining the relationship between food availability and adaptation.

Results show that while there is a negative and significant influence of cattle numbers on potholing, cattle numbers positively and significantly influence the growing of drought tolerant crops. This suggests that as cattle numbers increase, farmers are less likely to engage in potholing. Essentially, this underscores the importance of draught power for smallholder farmers to adapt to climate variability. Indeed, farmers reported that they recognize the significance of cattle ownership as a sign of wealth, particularly in that they get draught power from them.

It is further interesting to note that the location of farmers, that is, which country they are resident in, determines whether they will employ adaptation measures or not. There is a positive and significant relationship between country and adoption of conservation methods such as potholing, use of crop residues, conservation basin making and changing crops. This result suggests that farmers in Zambia are more inclined towards adaptation than those in Zimbabwe.

---

24 Consortium for Southern Africa's Food Emergency

CONCLUSION

This chapter has highlighted that while choice of adaptation strategies by small holder farmers may be inherent, it is largely influenced by both external and internal factors. It is therefore important to understand these factors in order to establish challenges to and opportunities for adaptation. In this regard, it is important to take note of those external factors, which can be addressed by implementing interventions that target either building on these factors if they are positive or finding mechanisms to address the constraints concerned. Activities such as training and encouraging social networks for farmers may go along way in strengthening education levels and social networks from which farmers draw in times of food insecurity induced by climate change.

CHAPTER NINE

ECOSYSTEM BASED APPROACHES
AND CLIMATE CHANGE ADAPTATION

9.0 Introduction

Adaptation, which is regarded as the adjustment in ecological, social or economic systems in response to actual or expected climatic stimuli and their effects, involves changes in processes, practices or structures to moderate or offset potential damages or to take advantage of opportunities associated with changes in climate. These adjustments intend to reduce the vulnerability of communities, regions or activities from climatic change and variability (Smit & Pilifosova 2003). There is evidence that higher levels of ecosystem diversity afford greater degrees of resilience to climate change impacts for both people and nature. A diversity of species and agricultural varieties, for example, means that farmers are able to adapt their agricultural practices in line with changing climatic conditions, thus reducing their vulnerability to climate-related shocks (IIED 2006).

It is therefore acknowledged that functionally diverse communities are more likely to adapt to climate change and climate variability than impoverished ones. This can have important implications for the designing of activities aimed at mitigating and adapting to climate change. Therefore, conservation of genotypes, species and functional types, along with the reduction of habitat loss, fragmentation and degradation, may promote the long-term persistence of ecosystems and the provision of ecosystem goods and services (IIED 2006). Therefore, much focus of adaptation strategies has been the use of ecosystem approaches such as the conservation of water resources, wetlands for both hydrological sustainability and human water supply; forest conservation for carbon sink and alternative source of energy such as the use of bio-fuels to reduce carbon emission (IIED 2006).

Adaptation measures have considered this ecosystem approach in a sustainable manner since ecosystem reflects dynamic complexities of plant, animal and micro-organism communities and their non-living environment interacting as a functional unit. The nature of these interactions can be direct and obvious or indirect and obscure. Any change in the constituents of an ecosystem can change the nature of these interactions with sometimes unexpected results (IIED 2006). These ecosystem alterations are inevitable under the threats of climate change. Therefore considering the use of ecosystem measures can promote stable resilience to both natural processes and community livelihoods.

9.1 Overview of Ecosystem Approaches as a response to climate change

Ecosystems play a significant role in regulating global climate. However, changes in biodiversity can affect this regulatory system. Forests, for example, are both a source and sink of carbon. Therefore, biodiversity management can play an important role in mitigating climate change through conservation of existing carbon sinks (e.g. forests and peat lands); regeneration of potential sinks (e.g. through afforestation and reforestation); and reduction of carbon emissions (e.g. through use of bio-fuels in place of fossil fuels) (IIED 2006).

Biodiversity can also make a contribution to climate change adaptation: mangrove systems, for example, are a highly effective natural flood control mechanism which will become increasingly important with sea level rise and the increasing frequency of extreme climatic events (IIED 2006). Hence, effective biodiversity conservation and management can lead to higher levels of carbon sequestration and hence climate change mitigation. For example, forest management activities such as increasing rotation age, low intensity harvesting, reduced impact logging, leaving woody debris, harvesting which emulates natural disturbance regimes, avoiding fragmentation, provision of buffer zones and natural fire regimes can simultaneously provide biodiversity and climate benefits (IIED 2004). This is also true for certain agro-forestry, revegetation, grassland management and agricultural practices such as recycling and use of organic materials (ibid).

Likewise, integrated watershed management can conserve watershed biodiversity in addition to increasing water retention and availability in times of drought; decreasing the chance of flash floods and maintaining vegetation as a carbon sink (IIED 2004). Conservation of biodiversity and maintenance of ecosystem integrity may be a key objective towards improving the adaptive capacity of such groups to cope with climate change. Functionally diverse systems may be better able to adapt to climate change and climate variability than functionally impoverished systems. A larger gene pool will facilitate the emergence of genotypes that are better adapted to changed climatic conditions. As biodiversity is lost, options for change are diminished and human society becomes more vulnerable (IIED 2004).

Ecological diversification plays an important role to managing climatic variability and change. In Mozambique, for example, people have plots on high ground for when there is a lot of rain and on low ground for when there is little rain; they also use lakes or depressions for cultivation in the aftermath of rains. People typically grow a diversity of crops and fruits in addition to cash crops and also use communal forest areas for grazing livestock and extracting forest products (Eriksen et al. 2008).

Therefore, diversification is a key strategy for coping with and adapting to the consequences of climate change. For example, in Uganda, fishers have diversified their strategies to adapt to fluctuations in lake levels and variations in fish productivity. In addition to fishing, they cultivate crops, maintain livestock, and collect firewood, reeds and papyrus from swamp areas and temporary migration and trade. They also reduce vulnerability by engaging in collective action for income-generating activities and access to markets, credit and relief. However, the poorest and weakest are often unable to gain membership in such groups and survive on casual employment, favours, gifts or urban migration (Eriksen *et al.* 2008).

Livestock herding, including nomadic pastoralism, remains one of the indigenous strategies best adapted to frequent droughts in dry land areas such as in Namibia and Botswana. Seasonal movement of livestock, splitting up of herds, changing herd composition and distributing livestock among relatives and friends in different areas minimize risk from droughts, floods or diseases. However, appropriation of the wetter areas and water sources for cultivation and fencing inhibit these crucial strategies (Eriksen *et al.* 2008).

Many in East Africa depend on forests for their livelihoods; adaptation measures may include afforestation programmes in degraded lands using more adaptive species, change in forest use and forest products to reduce tree felling, enhancement of forest seed banks, reduction of habitat fragmentation and conservation of migration corridors and buffer zones. Active community involvement in forest management is necessary to allow access to forest resources in a sustainable manner and improve local livelihood options (IIED 2005).

Interestingly, water management affects how vulnerable freshwater systems are to climate change. The IPCC has recognised the need for regional coordination in water management, particularly in international and shared basins and it recommends that international basin authorities should be strengthened and backed by robust legal frameworks. Better incorporation of current climate variability into water-related management would, however, make adaptation to future climate change easier. Adaptation in the water sector involves altering hydrological characteristics to suit human demands and altering demand to suit water availability. On the supply side, adaptation could include increasing storage by building dams and reservoirs, desalination or rainwater storage. On the demand side, measures could include improving water use efficiency and water recycling, reducing irrigation by changing cropping practices, importing agricultural products to irrigation areas and promoting local practices such as rainwater harvesting (IIED 2007).

9.2 Ecosystem Based Approaches and Sustainable Development

Due to the increased challenges of climate change impacts, there is a growing understanding of the possibilities to choose mitigation and adaptation options and their implementation such that there is no conflict with other dimensions of sustainable development; or, where trade-offs are inevitable, to allow a rational choice to be made. Hence, the rationale is to ensure that several strategies and measures that would enhance adaptive and mitigative capacities would also advance sustainable development. However, the sustainable development benefits of mitigation options vary within a sector and over regions (Sathaye *et al.* 2007).

For instance, to combat global warming, the UNFCCC was entered into following the Rio summit on Sustainable Development in 1992. The objective of UNFCCC (Article 2) is to stabilize greenhouse gas (GHG) concentration in the atmosphere at a level that would prevent dangerous anthropogenic interference with the climate system within a time frame sufficient enough to allow ecosystems to adapt naturally to climate change, to ensure that food production is not threatened and to ensure that economic development proceeds in a sustainable manner (Singh 2008).

The concerns about carbon-based conservation are underlain by the fact that because most terrestrial biomass carbon is in trees, silvicultural interventions that promote tree growth simultaneously promote carbon sequestration. This means that the sort of forest "improvement" treatments that foresters have long advocated for timber production are entirely appropriate for increasing forest-wide rates of carbon uptake. Unfortunately, it also means that efforts at maximizing forest carbon stocks share the same practical and philosophical limitations of striving to maximize timber yields (Ludwig *et al.* 1993; Elsevier 2009).

While there are many parallels between the forest carbon debate and the well publicized bio-fuel controversy (e.g., Koh &Ghazoul 2008), carbon-based conservation has received less scrutiny, partially because it involves protecting forests and planting trees, activities that most people endorse. To start on a positive note, protecting mature forests from degradation or destruction is an effective way to both reduce net emissions of greenhouse gases and to protect biodiversity (Venter *et al.* 2009).

Photo 29: Unsustainable harvesting of trees continues to cause depletion of forests in close proximity to human settlements.

While mature forests have relatively low incremental rates of net carbon uptake, with any positive "time value" of carbon (i.e., if the value of carbon sequestered in the future is discounted relative to the present), it never makes carbon-sense to allow any sort of anthropogenic disturbances (Harmon *et al.* 1990; Fargione *et al.* 2008). It follows that REDD funds should be used to expand and more effectively manage protected areas that contain mature forests, with appropriate safeguards for local use and compensation for lost access (Elsevier 2009).

Generally, the focus is to ensure that adaptation options that improve productivity of resource use, whether energy, water, or land, yield positive benefits across economic, social and environmental aspects as the key dimensions of sustainable development. However, other categories of mitigation options such as the use of bio-fuels plants e.g. Jatropha produced on productive land (probably suitable for staple food production); to mitigate carbon emission have a more uncertain impact and depend on the wider socioeconomic context within which the option is implemented (Sathaye *et al.* 2007).

Hence, it is not just climate change itself that can have an impact on natural systems, such as biodiversity. In some cases, the strategies that are adopted to mitigate climate change can affect biodiversity, both positively and negatively. Again, investment in renewable energy technology may provide climate change benefits but outcomes for biodiversity are often poor. For example, some bio-energy plantations replace sites with high biodiversity, introduce alien species and use damaging agro-chemicals. Large hydropower schemes can cause loss of terrestrial and aquatic biodiversity, inhibit fish migration and lead to mercury contamination (Montgomery *et al.* 2000). They can also be net emitters of greenhouse gases if submerged soils and vegetation decay and release Carbon dioxide and methane. By contrast, fuel wood conservation measures, such as efficient stoves and biogas use can conserve carbon reservoirs and reduce pressure on forests (IIED 2004).

For instance, reducing deforestation can have significant biodiversity, soil and water conservation benefits, but may result in loss of economic welfare for some stakeholders. Appropriately designed forestation and bio energy plantations can lead to reclamation of degraded land, manage water runoff, retain soil carbon and benefit rural economies but could compete with land for agriculture and may be negative for biodiversity (Sathaye *et al.* 2007; Collier *et al.* 2008).

The concept of becoming 'carbon neutral' is gaining popularity with many businesses that wish to contribute to climate change mitigation activities by offsetting their carbon emissions. Likewise, many nations have committed to reducing their net greenhouse gas emissions under the Kyoto Protocol of the UNFCCC. Projects designed to sequester carbon and hence mitigate climate change present opportunities to incorporate biodiversity considerations. Afforestation and reforestation activities can restore watershed functions, establish biological corridors and provide considerable biodiversity benefits if a variety of different aged native tree species are planted. Monocultures, however, not only reduce biodiversity but also increase the chances of pest attacks, thus challenging the permanence of carbon stocks. The location of afforestation and reforestation projects is also important. Replacing native grasslands, wetlands, shrub lands or heath lands may lead to dramatic biodiversity losses and also lower the relative increase in carbon sequestered compared to implementing such projects on degraded land (IIED 2004).

Generally, there are good possibilities for reinforcing sustainable development through mitigation actions in most sectors, but particularly in waste management, transportation and building sectors, notably through decreased energy use and reduced pollution (Sathaye *et al.* 2007). Likewise, climate-related policies, such as energy efficiency, are often economically beneficial, improve energy security and reduce local pollutant emissions. In most cases, energy supply mitigation options can also be designed to achieve other sustainable development benefits, such as avoided displacement of local populations, job creation and rationalized human settlements design (Sathaye *et al.* 2007).

Appropriately designed Community Based Forest Management policy can provide means to sustain and strengthen community livelihoods and at the same time avoid deforestation (see Photo 29), restore forest cover and density, provide carbon mitigation and create rural assets. This mechanism potentially has additional social, economic and ecological benefits. The international carbon market is a promising channel for improving livelihood opportunities for the rural poor in the forest areas (IUCN 2007) and can simultaneously strengthen existing national policies and programmes for forest protection and management (Singh 2008; Stern 2006).

9.3 Challenges for Synergies of Climate Change Responses and Sustainability

Globally, attention on climate change is predominantly focused not on adaptation but on carbon and methane emissions and mitigation. While there are some actions that Africans can take to reduce their emissions, particularly to do with land use and deforestation, by far the most important aspects of mitigation for Africa are the implications of the mitigation and adaptation strategies chosen by the rest of the world. At one extreme, some strategies for global mitigation have serious adverse consequences for Africa and so would be damaging for the region even if they succeeded in arresting global warming. At the other extreme, some strategies create new income-earning opportunities for Africa, although whether these opportunities are harnessed is again contingent upon human response (Collier *et al.* 2008).

Sustainable development has become part of all climate change policy discussions at the global level, particularly due to adoption of Agenda 21 and the various Conventions resulting from the UNCED-1992. The generally accepted and used definition as given by the Brundtland Commission is "development that meets the needs of the present without compromising the ability of future generations to meet their own needs" (WCED 1987).

Sustainable development has become an integrating concept embracing economic, social and environmental issues. Sustainable development does not preclude the use of exhaustible natural resources but requires that any use be appropriately offset. However, sustainable development cannot be achieved without significant economic growth in the developing countries (Goldenberg *et al.* 1995). Three critical components in promoting sustainable development are economic growth, social equity and environmental sustainability. Hence, the decline and degradation of natural resources such as land, soil, forests, biodiversity and groundwater resulting from current unsustainable use patterns are likely to be aggravated due to climate change in the next 25 to 50 years. Africa, South Asia and some regions of Latin America are already experiencing severe land degradation and freshwater scarcity problems (UNEP 1999).

Also, the diversion effect is already evident. One manifestation is consumer concern with 'food miles' and the potential damage that this could do to African suppliers of non-traditional agricultural exports. Another more important one is changes in relative prices, including the current high prices of cereals and many other food crops. While the main factors driving the higher prices are increased food consumption (particularly in Asia, with associated use of cereals as animal feed) and drought in some food-producing regions, the conversion of food-producing land to bio-fuels production is a further element. A significant 10% of world maize production now goes to bio-fuels production (Collier *et al.* 2008).

Since Africa is a net food importer, these price changes are having a significant negative effect on the terms of trade of many African countries. The fundamental problem here is not with bio-fuels production per se, but with the fact that production subsidies in some oil producing countries are causing inappropriate bio-fuels crops to be grown in the wrong places. In the USA, bio-fuels production displaces maize otherwise destined for food supply. In contrast, the opportunity cost of Brazilian bio-fuels production from sugar cane is much lower, while in Mozambique just 9% of arable land is at present cultivated and much of the rest is suitable for bio-fuels production from sugar cane or Jatropha with little or no impact on food production (Collier *et al.* 2008).

CONCLUSION

Generally, sustainable development can only be attained through sustainable human relations with nature. Climate change has a direct link with development concerns and therefore, there must be initiatives to promote sustainability through adaptation measures. On the other hand, there are many ways to pursue sustainable development strategies that contribute to mitigation and adaptation of climate change. A few examples include adoption of cost-effective energy-efficient technologies in electricity generation, transmission distribution and end-use can reduce costs and local pollution in addition to reduction of greenhouse gas emissions. A shift to renewables, some of which are already cost effective, can enhance sustainable energy supply and can reduce local pollution and greenhouse gas emissions. Adoption of forest conservation, reforestation, afforestation and sustainable forest management practices can also contribute to conservation of biodiversity, watershed protection, rural employment generation, increased incomes to forest dwellers and carbon sink enhancement (Sathaye 2006). Rational energy pricing based on long-run-marginal cost principle can level the playing field for renewables, increase the spread of energy-efficient and renewable energy technologies and the economic viability of utility companies, ultimately leading to greenhouse gas emission reduction (Sathaye *et al.* 2006).

CHAPTER TEN

BIOFUELS
Challenges and Opportunities

10.0 Introduction

In recent years, biofuels have rapidly emerged as a major issue for agricultural development, energy policy and natural resource management. The growing demand for biofuels is being driven by recent high oil prices, energy security concerns and global climate change. In Africa, there is growing interest from foreign private investors in establishing biofuel projects (see Photos 30 and 31 for such a project). For Tanzania, biofuel production has the potential to provide a substitute for costly oil imports (currently US$ 1.3-1.6 billion per year, 25% of total foreign exchange earnings). Biofuels also have the potential to provide a new source of agricultural income and economic growth in rural areas and a source of improvements in local infrastructure and broader development. Although many biofuel investments involve large plantations, biofuel production can also be carried out by smallholder farmers as well as

Photo 30: Sisal waste can efficiently be harvested to produce energy, particularly, biogas.

through 'out-grower' or local contracted farmer arrangements (Sulle & Nelson 2009; IIASA[25] 2009; Keeney & Nanninga 2008).

For African countries, this is leading to growing interest from Western and Asian private investors in biofuel projects, as well as growing support from bilateral and multilateral donors for incorporating biofuels into government policies and development plans. For countries in Africa which are non-oil producers, biofuel production has the potential to provide at least a partial substitute for costly oil imports, which are one of the major uses of foreign exchange and sources of inflation in African economies. Biofuel crops such as oils (palm, coconut, jatropha and sunflower) may provide important new opportunities for improving the returns from agriculture, including on relatively unproductive or infertile lands (Sulle & Nelson 2009).

10.1 Implications of bio-fuels production for biodiversity resources

External interest in biofuel production in African countries is driven largely by the low cost of land and labour in rural Africa (Cotula *et al.* 2008). Investors are targeting many areas of land which are perceived as being 'unused' or 'marginal' in terms of their productivity and agricultural potential. With interest in allocating such areas for biofuel increasing, the security of land tenure and access or use rights on the part of local resident communities across rural African landscapes is potentially at risk (Sulle & Nelson 2009).

Moreover, the spread of biofuels in different parts of the world has also raised concerns from civil society organisations, local communities and other parties. This concern has been attributed to the fact that the environmental impact of biofuel plantations could involve water scarcity and deforestation, particularly in coastal areas (Sulle & Nelson 2009). Considerable concern has been expressed about the impacts of biofuel development in terms of environment and biodiversity outcomes, food security locally and nationally and local access and rights over land (Kamanga 2008; Gordon-Maclean *et al.* 2008). Some of the actual and potential agronomic and ecological threats include impacts on the soil and environment. For example, biofuel plantations that involve the clearing of areas with high levels of biodiversity or that replace natural habitats such as *Miombo* woodlands; large biofuel plantations that can block wildlife migratory routes in parts of the country, especially in areas surrounding or near to wildlife conservation areas (Sulle & Nelson 2009).

---

25 IIASA: International Institute for Applied Systems Analysis

The reduction in global biodiversity has emerged as one of the greatest environmental threats of the 21st century due to climate change and climate change mitigation strategies like the use of bio energy for reducing carbon emission. Urban and subsistence agricultural developments have traditionally been primary drivers of encroachment on important, biodiversity-sustaining ecosystems. But a new agricultural trend, the use of plant biomass to provide liquid fuels, is exacerbating agriculture's impact on biodiversity. These fuels, called biofuels, are changing land-use patterns in many regions around the world, including some of the most diverse and sensitive regions on the planet (Keeney & Nanninga 2008).

The first pathway for biodiversity loss is habitat loss following land conversion for crop production, for example from forest or grassland. As the CBD (2008) notes, many current biofuel crops are well suited for tropical areas. This increases the economic incentives in countries with biofuel production potential to convert natural ecosystems into feedstock plantations (e.g. oil palm), causing a loss of wild biodiversity in these areas. While oil palm plantations do not need much fertilizer or pesticide, even on poor soils, their expansion can lead to loss of rainforests (FAO 2008).

Although loss of natural habitats through land conversion for biofuel feedstock production has been reported in some countries (Curran *et al.* 2004; Soyka, Palmer & Engel 2007), the data and analysis needed to assess its extent and consequences are still lacking. Nelson & Robertson (2008) as cited in FAO (2008) examined how rising commodity prices caused by increased biofuel demand could induce land-use change and intensification in Brazil and found that agricultural expansion driven by higher prices could endanger areas rich in bird species diversity (FAO 2008).

The second major pathway is loss of agro biodiversity, induced by intensification on croplands, in the form of crop genetic uniformity. Most biofuel feedstock plantations are based on a single species. There are also concerns about low levels of genetic diversity in grasses used as feed stocks, such as sugar cane, which increases the susceptibility of these crops to new pests and diseases. Conversely, the reverse is true for a crop such as jatropha, which possesses an extremely high degree of genetic diversity, most of which is unimproved, resulting in a broad range of genetic characteristics that undermine its commercial value (FAO 2008).

When forests or grasslands are converted to farmland, be it to produce biofuel feed stocks or to produce other crops displaced by feedstock production, carbon stored in the soil is released into the atmosphere. The effects can be so great that they negate the benefits of biofuels and lead to a net increase in greenhouse gas emissions when replacing fossil fuels (FAO 2008). Biofuel production can affect habitat for biodiversity. For instance, habitat is lost when natural landscapes are converted into energy-crop plantations or peat lands are drained. In some instances, however, biofuel crops can have a positive impact, for instance when they are used to restore degraded lands. In order to ensure an environmentally sustainable biofuel production, it is important that good agricultural practices be observed and measures to ensure sustainability should be applied consistently to all crops. Moreover, national policies will need to recognise the international consequences of biofuel development (FAO 2008).

Usually, a difference can be made between direct and indirect impacts of biofuels on biodiversity. However, there exists much vagueness with regard to the boundary line between direct and indirect impacts. In relation to impacts on biodiversity: indirect impacts mostly refer to saving species by climate change mitigation (e.g. biofuels decrease carbon emissions and thus reduce climate change and climate change therefore has a reduced consequential impact on biodiversity); whereas direct impacts refer to interferences in ecosystems (e.g. the direct removal of existing high biodiversity value forests for palm oil plantations).

Photo 31: Biofuels decrease carbon emmissions reducing climate change leading to reduced consequential impact on biodiversity.

From this point of view, the impacts described here mainly reflect direct effects. Although indirect impacts certainly deserve attention, there has already been an emphasis in the current debate on indirect impacts while direct impacts have received less consideration (Biemans *et al.* 2008).

Until recently, many policy-makers assumed that the replacement of fossil fuels with fuels generated from biomass would have significant and positive climate-change effects by generating lower levels of the greenhouse gases that contribute to global warming. Bio-energy crops can reduce or offset greenhouse gas emissions by directly removing carbon dioxide from the air as they grow and storing it in crop biomass and soil. In addition to biofuels, many of these crops generate co-products such as protein for animal feed, thus saving on energy that would have been used to make feed by other means (FAO 2008).

However, despite these potential benefits, scientific studies have revealed that different biofuels vary widely in their greenhouse gas balances when compared with petrol. Depending on the methods used to produce the feedstock and process the fuel, some crops can even generate more greenhouse gases than do fossil fuels. For example, nitrous oxide, a greenhouse gas with a global- warming potential around 300 times greater than that of carbon dioxide, is released from nitrogen fertilizers. Moreover, greenhouse gases are emitted at other stages in the production of bio-energy crops and biofuels: in producing the fertilizers, pesticides and fuel used in farming, during chemical processing, transport and distribution up to the final use (FAO 2008).

10.2 Ensuring environmentally sustainable biofuel production

10.2.1 Good practices
Good practices aim to apply available knowledge to address the sustainability dimensions of on-farm biofuel feedstock production, harvesting and processing. This aim applies to natural-resource management issues such as land, soil, water and biodiversity as well as to the life-cycle analysis used to estimate greenhouse gas emissions and determine whether a specific biofuel is more climate-change friendly than a fossil fuel. In practical terms, soil, water and crop protection; energy and water management; nutrient and agrochemical management; biodiversity and landscape conservation; harvesting, processing and distribution all count among the areas where good practices are needed to address sustainable bio-energy development (FAO 2008).

10.2.2 Promoting organic agriculture
Conservation agriculture is one practice that sets out to achieve sustainable and profitable agriculture for farmers and rural people by employing minimum soil

disturbance, permanent organic soil cover and diversified crop rotations. In the context of the current focus on carbon storage and on technologies that reduce energy intensity it seems especially appropriate. The approach also proves responsive to situations where labour is scarce and there is a need to conserve soil moisture and fertility. Interventions such as mechanical soil tillage are reduced to a minimum and inputs such as agrochemicals and nutrients of mineral or organic origin are applied at an optimum level and in amounts that do not disrupt biological processes. Conservation agriculture has been shown to be effective across a variety of agro-ecological zones and farming systems. Good farming practices coupled with good forestry practices could greatly reduce the environmental costs associated with the possible promotion of sustainable intensification at forest margins. Approaches based on agro-silvo-pasture-livestock integration could be considered also when bio-energy crops form part of the mix (FAO 2008).

10.2.3 Standard, sustainability criteria and compliance

Although the multiple and diverse environmental impacts of bio-energy development do not differ substantively from those of other forms of agriculture, the question remains of how they can best be assessed and reflected in field activities. Existing environmental impact-assessment techniques and strategic environmental assessments offer a good starting point for analysing the biophysical factors. There also exists a wealth of technical knowledge drawn from agricultural development during the past 60 years. New contributions from the bio-energy context include analytical frameworks for bio-energy and food security and for bio-energy impact analysis (FAO forthcoming (a) and (b)); work on the aggregate environmental impacts, including soil acidification, excessive fertilizer use, biodiversity loss, air pollution and pesticide toxicity (Zah *et al.* 2007); and work on social and environmental sustainability criteria, including limits on deforestation, competition with food production, adverse impacts on biodiversity, soil erosion and nutrient leaching (FAO 2008).

CHAPTER ELEVEN

CONCLUSIONS AND RECOMMENDATIONS

There is concern that the world's poor in the Sub-Saharan Africa zone contributing the least to climate change are the most vulnerable and unable to adapt to the rising frequency and severity of extreme weather and climate variability (e.g., droughts and floods)—which have profound impacts on water supply, river flows, crop and fishery yields and health (Mirza 2003; Lamarre & Besancot 2008; Srinivasan *et al.* 2008). A warmer climate, with increased climate variability, is increasing the risk of both floods and droughts (Wetherald & Manabe 2002; IPCC 2007; Kundzewicz *et al.* 2007).

As there are a number of climatic and non-climatic drivers influencing flood and drought impacts, the realisation of risks depends on several factors. Climate-related trends of some components during the last decades have already been observed. For a number of components, for example groundwater, the lack of data makes it impossible to determine whether their state has changed in the recent past due to climate change. On the other hand, during recent decades, non-climatic drivers have exerted strong pressure on freshwater systems. This has resulted in water pollution, damming of rivers, wetland drainage, reduction in stream flow and lowering of the groundwater table (mainly due to irrigation). In comparison, climate-related changes have been small, although this is likely to be different in the future as the climate change signal becomes more evident (Kundzewicz *et al.* 2007).

Generally, the strong trends in climate change are already evident, the likelihood of further changes occurring and the increasing scale of potential climate impacts give urgency to addressing socio-economic adaptation more coherently (Easterling 2007). Although the ability to adapt and cope with climate change impacts is a function of wealth, scientific and technical knowledge, information, skills, infrastructure, institutions and equity; little has been done to integrate those adaptation necessities in policy and development initiatives among East and Southern African countries. The general thesis is that countries with limited economic resources, low level of technology, poor information and skills, poor infrastructure, unstable or weak institutions and inequitable empowerment and access to resources have little capacity to adapt and are highly vulnerable (IDRC/CCAA 2009).

Hence, deliberate measures planned ahead of time at local, regional, national, and international levels are needed to facilitate a broader range of responses. Many options for policy-based adaptation to climate change have been identified for agriculture, forests and fisheries. These can involve adaptation activities such as developing infrastructure, capacity building in the broader user community and institutions and in general, modifications to the decision-making environment under which management-level adaptation activities typically occur. However, in some countries in the regions highlighted in this book, , the process of 'mainstreaming' adaptation into policy planning in the face of risk and vulnerability at large is an important component of adaptation planning. However, there are formidable environmental, economic, informational, social, attitudinal, and behavioural barriers to the implementation of adaptation (Easterling 2007). Thus, prospects for successful adaptation to climate change will remain limited as long as factors (e.g., population growth, poverty and globalisation) that contribute to chronic vulnerability to, for example, drought and floods, are not resolved.

Adapting to current climate variability and future climate changes requires the development of systems that are capable of absorbing the current climate shocks and at the same time integrate future climate change risks in the society, economic and environment systems to ensure sustainability of the social, economic and environment systems (IDRC/CCAA 2009). These may include among others developing sustainable development policies; reducing population pressure on resources and environmental degradation; fostering community-based development and resource management initiatives. Other factors include maintaining the role of traditional knowledge and practices in adaptation; promoting education and awareness creation to enhance adaptation and promoting environmentally sound technologies. The strengthening of the flood control systems; promoting of crop diversity and agro-forestry systems and improving the early warning systems including the use of technical and traditional warning systems have also been emphasised (IDRC/CCAA 2009).

Moreover, a response to climate change will be more effective if it is organized globally and when it involves international understanding and collaboration (Hepburn & Stern 2008). Adaptation strategies should be integrated within the national development plans by ensuring that all possible vulnerabilities are consistent with it. There is need to prepare for the climate related shocks by relating them to how they can affect communities. This can only be appropriately done at the design of the development plans. However, it was observed that planning for adaptation is not that simple and needs not be undertaken in isolation but integrated across all sectors such that the resources required can be well planned for (IDRC/CCAA 2009; Majule 2009; Orindi & Murray 2005).

Although floods and droughts are natural events, the impacts associated with them can be reduced or controlled. However, in East Africa, like some other parts of the world including India and South Africa (Dyer 1981), there are prospects for out-of-season rainfall compensation of the deficit conditions. Accurate seasonal to inter-annual climate monitoring and forecasting could therefore contribute to improved planning and the management of climate sensitive activities, involving agricultural and water resources, hydroelectric power supply and tourism, among others (IDRC/CCAA 2009).

Despite evidence showing that African farmers have a low capacity to adapt to climate changes, they have, however, survived and coped in various ways over time. Better understanding of how they have done this is essential for designing incentives to enhance private adaptation. Supporting the coping strategies of local farmers through appropriate public policy and investment and collective actions can help increase the adoption of adaptation measures that will reduce the negative consequences of predicted changes in future climate, with great benefits to vulnerable rural communities in Africa (Hassan & Nhemachena 2008). Therefore, in the planning process for strategies based on what is available; one needs to have proper policies that will ensure sustainable adaptation. For countries sharing transboundary water resources, their interests need to be harmonized (Majule 2009).

Hence, traditional knowledge of local communities represents an important, yet currently largely under-used resource for adapting to climate change impacts (Huntington & Fox 2005). Empirical knowledge from past experience in dealing with climate-related natural disasters such as droughts and floods (Osman-Elasha et al. 2006), health crises (Wandiga et al. 2006), as well as longer term trends in mean conditions (Huntington & Fox 2005; McCarthy & Martello 2005), can be particularly helpful in understanding the coping strategies and adaptive capacity of indigenous and other communities relying on oral traditions (Carter et al. 2007). Experience gained in dealing with climate-related natural disasters and documentation and education using both modern methods and traditional knowledge can assist in understanding the coping strategies and adaptive capacity of vulnerable communities and in defining critical thresholds of impact to be avoided (Carter et al. 2007; Hassan & Nhemachena 2008).

While the importance of indigenous knowledge has been realized in the design and implementation of sustainable development projects, little has been done to incorporate this into formal climate change mitigation and adaptation strategies. Climate change cannot be divorced from sustainable development as sustainable development may be the most effective way to frame the mitigation question and a crucial dimension of climate change adaptation and impacts (Swart *et al.* 2003; Cohen *et al.* 1998). Incorporating indigenous knowledge into climate change policies can lead to the development of effective mitigation and adaptation strategies that are cost-effective, participatory and sustainable (Robinson & Herbert 2001; Hunn 1993). However, incorporating indigenous knowledge into climate change concerns should not be done at the expense of modern/ western scientific knowledge. Indigenous knowledge should complement, rather than compete with global knowledge systems (Nyong *et al.* 2007).

There is no doubt that extreme climatic events such as drought, as well as major biotic problems may have a major impact on the availability of, and access to, seed. Therefore, strengthening formal and informal seed systems is an important adaptive strategy. Local seed systems can be very resilient and continue to function following severe climatic or other disasters (Sperling et al. 2004). Initiatives such as Direct Seed Distribution (DSD) or Seeds and Tools (S&T) have been a common and well publicised response to disasters (climatic or otherwise) (Sperling et al. 2004). Indeed, very often it is not seed per se that is the problem, but access to seed – access that is constrained by factors such as livelihood and interactions with local markets (Challinor et al. 2007).

In general, emerging literature shows that the distribution of adaptive capacity within and across societies represents a major challenge for development and a major constraint to the effectiveness of any adaptation strategy. Some adaptations that address changing economic and social conditions may increase vulnerability to climate change, just as adaptations to climate change may increase vulnerability to other changes (Adger *et al.* 2007). The existence of these barriers does not mean that the development community does not recognise the linkage between development and climate-change adaptation.

Climate change is identified as a serious risk to poverty reduction in developing countries, particularly because these countries have limited capacity to cope with current climate variability and extremes not to mention future climate change (Schipper & Pelling 2006). Adaptation measures will need to be integrated into strategies of poverty reduction to ensure sustainable development and this will require improved governance, mainstreaming of climate-change measures and the integration of climate-change impacts information into national economic projections (AfDB *et al.* 2003; Davidson *et al.* 2003; Orindi & Murray 2005). The Commission for Africa (2005) explicitly links the need to address climate-change risks with achievement of poverty reduction and sustainable growth (Yohe *et al.* 2007).

REFERENCES

Abbot, J., S. Neba, and M. Khen. 1999. *Turning Our Eyes from the Forest.* The role of the Livelihoods Programme at Kilum-Ijim Forest Project, Cameroon in changing attitudes and behaviour towards forest use and conservation. NP:NP.

Adams MA, et al (2004) "Biodiversity Conservation and the Eradication of Poverty". *Science Journal*: Vol. 306: No. 5699: 1146–1149.

Adams W. 2004. Against Extinction. London: Earthscan.

Adams, W. M. and Hulme, D.(2001) *If community conservation is the answer in Africa, what is the question?* Oryx, 35 (3), 193–200, London: Fontana.

Aday, L.A., Begley, C.E., Lairson, D.R., Slater, C.H., Richard, A.J., Montoya, I.D., "1999. A framework for assessing the effectiveness, efficiency, and equity of behavioral healthcare." American Journal of Managed Care 5, 25–44.

Adger, W.N (2001) Scales of Governance and Environmental Justice for Adaptation and Adaptation of Climate Change: Journal of International Development; Vol. 1: 921-931.

Adger, W.N et al (2005) Successful Adaptation to Climate Change Across Scales: Global Environmental Change Vol.15: 77-86 UK-Elsevier.

Adger, W.N. et al (2007) Assessment of Adaptation Practices, Options, Constraints and Capacity. Climate Change 2007: Impacts, Adaptation and Vulnerability. Contribution of Working Group II to the Fourth Assessment Report of the Intergovernmental Panel on Climate Change In Parry, M.L. et al (Eds.) Cambridge, Cambridge, University Press, pp 717-743.

Agrawal, A. (1999) "'Community'-in-conservation: tracing the outlines of an enchanting concept", in R. Jeffery and N. Sundar, eds, *A New Moral Economy for India's Forests? Discourses of 'Community' and Participation,* Delhi. Sage Publications. , 92–108.

Agrawal, A. and Gibson, C. C. (1999) Enchantment and disenchantment: the role of 'community' in natural resource conservation, World Development, 27 (4), 629–649.

Agrawala, S. Et al (2003). Development and Climate Change in Bangladesh: Focus on Coastal Flooding and the Sundarban. Working Party on Global and Structural Policies, Working Party on Development Cooperation and Environment: Organization for Economic Cooperation and Development (OECD).

Amanor, K. (2003) Natural and Cultural Assets and Participatory Forest Management in West Africa: International Conference on Natural Assets. Conference Paper Series No. 8 New York: Sons, 320 pp.

Anderson K, and A. Bows (2005). "Reframing the Climate Change Challenge in Light of Post 2000 Emission Trends", in *Philosophical Transactions of the Royal Society A: Mathematical, Physical and Engineering Sciences* (2008) 366: 3863–3882. NP: NP

Apsey, M., Laishly, D., Nordin, V., Paille, G., 2000. "The perpetual forest: using lessons from the past to sustain Canada's forests in the future", *Forestry Chronicle* 76 (6), 29– 53.

Arnon, B. (2006). "On elephants, giraffes, and social development: *Oxford-Community Development Journal*; Vol. 41: No. 3: 367-380

Arnell, N.W., (2004). Climate Change and Global Water Resources: SRES Emissions and Socio-Economic Scenarios. *Global Environmental Change* Vol. 14: 31–52.

Assan J.K et al (2009) "Environmental Variability and Vulnerable Livelihoods: Minimising Risks and Optimising Opportunities for Poverty Alleviation;" *journal of International Development* Vol. 21: 403-418.

Assan JK. (2008). "Generational Differences in Internal Migration: Derelict Economies, Exploitative Employment and Livelihood Discontent." *In International Development Planning Review* Vol. 30 No. 4: 377– 398.

Assan, J.K and Kumar, P (2009) Livelihood Options for the Poor in the Changing Environment: Journal of International Development Vol. 21: 393-402.

Baker R.H.A et al (2000). "The Role of Climatic Mapping in Predicting the Potential Geographical Distribution of Non-Indigenous Pests Under Current and Future Climates". *In Agriculture, Ecosystems and Environment* 82: 57–71.

Biemans, M; Waarts, Y; Nieto, A; Goba, V; Jones-Walters, L and Zöckler, C (2008). *Impacts of Biofuel Production on Biodiversity in Europe.* The Nertherlands-European Centre for Nature Conservation

Beierle, T.C., Konisky, D.M. (2000). "Values, Conflict, and Trust in Participatory Environmental Planning", *Journal of Policy Analysis and Management* 19: 587–602.

Bell, R. (1999) CBNRM and Other Acronyms: An Overview and Challenges in the Southern African Region. Paper presented at the CASS/ PLAAS Inaugural Meeting on Community Based Natural Resource Management in Southern Africa: A Regional Programme of Analysis and Communication, Harare, 21–23 September.

Biega Landscape (2010) *Democratic Republic of Congo*, NP: Conservation International.

Binot, A et al (2009) "The Challenges of Participatory Natural Resources Management with Mobile Hearders at the Scale of Sub-Saharan African Protected Areas", *Biodiversity Conservation Journal* Vol. 18: 2645-2662.

Blennow, K and Persson, J (2009) "Climate change: Motivation for taking measure to adapt", *Global Environmental Change* Vol. 1: 100-104.

Bocoum, A. (2000). "Natural forest in Segue, Koro-Mopti Circle, Mali", In *FAO 2000a*.

Bohle, H (2001) *Vulnerability and Criticality: Perspectives from Social Geography*. IHDP.

Boko, M. I. et al (2007) "Africa in Climate Change", In Parry, M.L. et al (Eds) 2007 *Impacts, Adaptation and Vulnerability. Contribution of Working Group II to the Fourth* Assessment Report of the Intergovernmental Panel on Climate Change. Cambridge University Press: 433-467.

Bond, I; M. Grieg-Gran; S. Wertz-Kanounnikoff, P. Hazlewood; S. Wunder and A. Angelsen (2009) *Incentives to Sustain Forest Ecosystem Services;* NP: International Institute for Environment and Development.

Boote K.J and J.W Jones (1998). "Simulation of Crop Growth: CROPGRO model", In: *Agricultural Systems Modelling and Simulation*—Peart R.M and R.B Curry (Eds). (1998) New York: Marcel Dekker: 651–692.

Borrini-Feyerabend, G., Buchan, D., 1997. *Beyond fences: Seeking social sustainability in conservation*. International Union for the Conservation of Nature (IUCN), Gland. 2 Vols, 283 pp.

Borrini-Feyerabend, G., Farvar, M. T., Nguinguiri, J. C. and Ndangang, V. A. (2000) *Co-management of Natural Resources: Organising, Negotiating and Learning-by-Doing*, Kasparek Verlag, Heidelberg GTZ and IUNC, .

Bosetti, V et al (2009) "Climate Policy after 2012: *Economic Studies* Vol. 55 No. 2: 235-254.

Botes, L., Van Rensburg, D., 2000. "Community participation in development: nine plagues and twelve commandments." *Community Development* Journal 35, 41–58.

Braga, C., (2001). "They're squeezing us!: matrilineal kinship, power and agricultural policies: case study of Issa Malanga, Niassa Province. In:Waterhouse, R., Vijhhuizen, C. (Eds.), *Strategic women, gainful men: gender, land and natural resources in different rural context in Mozambique*. Nucleo Maputo de Estudos de Terra and University of Eduardo Mondlane, 199–226.

Bray, D.B., et al (2002). "Mexico's community-managed forests as a global model for sustainable landscapes", *Conservation Biology* 17, 672–677.

Bradshaw et al. (2004) "Farm-Level Adaptation to Climatic Variability and Change: Crop Diversification in the Canadian Prairies", *Climatic Change* Vol. 67: 119–141.

Brosius, J. P., Tsing, A. L. and Zerner, C. (1998) "Representing communities: histories and politics of community-based natural resource management", *Society and Natural Resources,* 11 (2), 157–168.

Brown, David (1999) *Principles and Practice of Forest Co-Management: Evidence from West- Central Africa. European Union Tropical Forestry* Paper 2. London: ODI, and Brussels: European Commission.

Brown, O. and Crawford A (2008) Assessing the Security Implication of Climate Change for West Africa, Winnipeg, Manitoba: IISD.

Bührs, T., Aplin, G., 1999. "Pathways towards sustainability: the Australian approach". Journal of *Environmental Planning and Management* 42: 315–340.

Burke, M.B. et al (2009) "Shifts in African Crop Climates by 2050, and the Implications for Crop Improvement and Genetic Resource Conservation". Journal of *Global Environmental Change,* Vol. 699; Sweden, Elsevier.

Burton, I. et al. (2003). Livelihoods and Climate Change. Combining Disaster Risk Reduction, Natural Resource Management and Climate Change Adaptation in a New Approach to the Reduction of Vulnerability and Poverty. A Conceptual Framework Paper Prepared by the Task Force on Climate Change, Vulnerable Communities and Adaptation, IUCN/SEI/ IISD/Intercooperation.

Burke, M.B et al (2009) "Shifts in African Crop Climates by 2050, and the Implications for Crop Improvement and Genetic Resource Conservation": *Global Environmental Change* Vol. 19: 317-325.

Bush, M. 2000. "Learning by Doing in Co-Management for the Classified Forests of Guinea-Conakry". In *CM News* No. 4 September 2000, Gland.

Campbell, B et al (2001). Challenges to proponents of common property resource systems: despairing voices from the social forests of Zimbabwe. World Development 29 (4), 589– 600.

Campbell, B., N. Byron, P. Hobane, P., F. Matose, F. Madzudzo & E. Alden Wily. 1999. "Moving to Local Control of Woodland Resources – Can CAMPFIRE Go Beyond the Mega-Fauna?" *Society and Natural Resources,* 12, 501-509.

Canadell, Joseph G and Michael R. Raupach (2008) "Managing Forests for Climate Change Mitigation". *Science Journal* Vol. 320 No. 5882:1456-57.

Carley, M., (1994). Policy management for systems and methods of analysis for sustainable agriculture and rural development. London: IIED, 64 pp.

Castro, A.P., Nielsen, E., 2001. "Indigenous people and co-management: implications for conflict management". *Environmental Science and Policy* 4, 229–239.

Carr, E.R (2008) "Between Structure and Agency: Livelihoods and Adaptation in Ghana's Central Region": *Global Environmental Change* Vol. 18: 689-699.

CBD (Convention on Biological Diversity). 2008. *The potential impact of biofuels on biodiversity.* Note by the Executive Secretary for the Conference of the Parties to the Convention on Biological Diversity, 19–30 May 2008, Bonn, Germany (draft, 7 February 2008).

Challinor A. J., et al (2006). "Assessing the Vulnerability of Crop Productivity to Climate Change Thresholds Using an Integrated Crop-climate Model". In: Schellnhuber J. (Ed). (2006) *Avoiding Dangerous Climate Change.* Cambridge: Cambridge University Press, 187–94.

Challinor A.J (2005c) "Probabilistic Hindcasts of Crop Yield Over Western India". *Tellus* 57A: 498–512.

Challinor A.J and T.R, Wheeler (2007). "Crop Yield Reduction in the Tropics under Climate Change: Processes and Uncertainties". *Agricultural and Forest Meteorology;* Vol. 148: 343–356. Change at die Farm Level." In H. Kaiser and T. Drennen, (Eds). *Agricultural Dimensions of Global.*

Challinor A.J et al (2009b) *Methods, Skills and Resources for Climate Impacts Research.* Bulletin of the American Meteorological Society in press.

Challinor A.J, et al (2005d). "Simulation of the Impact of High Temperature Stress on Annual Crop Yields". *Agricultural and Forest Meteorology;* Vol. 135: 180–189.

Challinor, A et al (2007) "Asseing the Vulnerability of Food Crop Systems in Africa to Climate", *Climate Change Journal* Vol.83, No.3: 381-399, Netherlands-Springer.

Challinor, A.J et al (2009) "Crops and Climate Change: Progress, Trends, and Challenges in Simulating Impacts and Informing Adaptations", Vol. 60 No. 10: 2775-2789 In *Journal of Experimental Botany,* London-Oxford: University Press.

Chandler, W; R. Schaeffer; Z. Dadi; P.R shukla; F.Tudela; D. Daivdson; S. Alpan-Atamer (2009) Climate Change Mitigation in Developing Countries. Global Climate Change Proceedings.

Chhibber, A and L. Rachid (2008) "Disasters, Climate Change and Economic Development in Sub-Saharan Africa: Lessons and Directions"; Journal of *African Economies* Vol.17, Supplement 2: London: Oxford University Press.

Cochrane, K; C. De Young; D. Soto; T. Bahri (2009) (Eds) *Climate Change Implications for Fisheries and Aquaculture: Overview of the Current Scientific Knowledge.* N.P: Food and Agriculture Organisation.

Collier, P et al (2008) "Climate Change and Africa", *Oxford Review of Economic Policy* Vol. 24 No. 2: 337-353; London-Oxford University Press.

Collier, P et al (2008) "Vulnerability of Horticultural Crop Production to Extreme Weather Events", *Aspects of Applied Biology;* Vol. 88: 3–13.

Conroy, C., Mishra, A., Rai, A., 2002. "Learning from self-initiated community forest management in Orissa, India", *Forest Policy and Economics* 4 (3), 227–237.

Cooper, P. (2004) *Coping with Climatic Variability and Adapting to Climate Change: Rural Water Management in Dry-land Areas.* London: International Development Research Centre.

Cotula, L., Dyer, N. and Vermeulen, S. 2008. *Fuelling Exclusion? The Biofuels Boom and Poor People's Access to Land.* London: FAO and IIED.

Crabbe, M.J (2009) "Climate Change and Tropical Marine Agriculture", Journal *of Experimental Botany* Vol. 60 No.10: 2839-2844.

Craufurd, P.Q and T.R. Wheeler (2009) "Climate Change and the Flowering Time of Annual Crops", *Journal of Experimental Botany.* Vol. 60, No.9: 2529-2539: UK-University of Reading.

Croll, E. and Parkin, D., eds (1992) *Bush Base/Forest Farm: Culture, Environment and Development,* London: Routledge.

Davies et al (2008) *Climate Change Adaptation, Disaster Risk Reduction and Social Protection:* Sussex: Institute of Development Studies: United Kingdom-University of Sussex.

Davies, M et al (2009) Climate Change Adaptation, Disaster Risk Reduction and Social Protection: Complementary Roles in Agriculture and Rural Growth? IDS Working Paper No. 320. Centre for Social Protection and Climate Change and Development Centre.

Department of Water Affairs and Forestry (DWAF), 2003. DWAF/Danida PFM project: terms of reference for the development of PFM policy and Strategy. Department of Water Affairs and Forestry, Pretoria.

Dercon, S (2002) "Income Risk, Coping Strategies and Safety Nets", *The World Bank Research Observer*, Vol. 17, No. 2: 141-166.

Deressa, T.T and R.M, Hassan (2009) "Economic Impact of Climate Change on Crop Production in Ethiopia: Evidence From Cross-Section Measures", *Journal of African Economies*. Vol. 18, No. 04: 524-554, London: Oxford University Press.

Deressa, T.T et al (2009) "Determinants of Farmers' Choice of Adaptation Methods to Climate Change in the Nile Basin of Ethiopia", *Global Environmental Change* Vol. 19: 248-255.

Dessai, S (2005) "On the Role of Climate Scenarios for Adaptation Planning", *Global Environmental Change Part A*; Vol.15 No. 2: 87-97. Elsevier-Stockholm.

Dessai, S et al (2007) "On the Role of Climate Scenarios for Adaptation Planning", *Global Environmental Change, Part A*: Vol.15, No. 2: 87-97. Sweden-Elsevier.

Diaz S (1993). "Evidence of a Feedback Mechanism Limiting Plant Response to Elevated Carbon Dioxide", *Nature* (1993) 364: 616–617.

Dixon, R.K. et al (2003) "Life on the Edge: Vulnerability and Adaptation of African Ecosystems to Global Climate Change", *Mitigation and Adaptation Strategies for Global Change* Vol. 8: 93-113, Netherlands, Kluwer Academic Publishers.

Downing, T.E (2003). "Lessons from Famine Early Warning and Food Security for Understanding Adaptation to Climate Change: Toward a Vulnerability/Adaptation Science". In: Smith, J.B et al (2003) (Eds), *Climate Change Adaptive Capacity and Development*, London Imperial College Press.

Downing, T.E. (2002) 'Protecting the Vulnerable: Climate Change and Food Security', in J.C. Briden and T.E. Downing (Eds.) *Managing the Earth: The Linacre Lectures*, Oxford: Oxford University Press.

Dryzek, J.S., 1997. *The politics of the earth: environmental discourses.* Oxford: Oxford University Press, 240 pp.

Dubois, O. & J. Lowore. 2000. *The Journey towards Collaborative Forest Management in Africa:.* N.P: N.P.

Easterling W.E, et al (2007) Food, Fibre and Forest Products: In Parry, M.L et al 2007 (Eds) Climate Change: Impacts Adaptation and Vulnerability. Contribution of Working Group II to the Fourth Assessment Report on the Intergovernmental Panel on Climate Change; Cambridge, UK: Cambridge University Press. 273–313; Economics Vol. 75: 387-98.

Easterling, W (2007) Crop and Pasture Response to Climate Change: Pennsylvania: -Pennsylvania State University-University Park.

Easterling, W (Ed) 2007 *Adapting Agriculture to Climate Change:* N.P: Common Wealth Scientific and Industrial Research Organisation, Sustainable Ecosystems.

Ellis, F and G. Bahiigwa (2003) "Livelihoods and Rural Poverty Reduction in Uganda". *World Development* Vol. 31, No. 6: 997-1013; *Sweden-Elsevier*.

Els, H., Bothma, J.P., 2000. "Developing partnerships in a paradigm shift to achieve conservation reality in South Africa", *Koedoe* 43, 19–26.

Elsevier (Ed) 2005 "Introduction to Papers on Mitigation and Adaptation Strategies for Climate Change: Protecting Nature from Society or Protecting Society from Nature?" *Environmental Science & Policy* Vol. 8: 537–540.

Elsevier (Ed) 2009 "Dangers of Carbon-Based Conservation", *Global Environmental Change* Vol: 19 400-401

Enfors, E.I and L.J Gordon (2008) "Dealing with Drought: The challenge of using water system technologies to break dry land poverty traps". *Global Environmental Change* Vol. 18: 607-616.

Enters, T., Anderson, J., 1999. "Rethinking the decentralisation and devolution of biodiversity conservation". *Unasylva* 50, 6–11.

Enkvist, P., T. Naucler and J. Rosander (2007). *A cost curve for greenhouse gas reduction.* N.P: The McKinsey.

Eriksen, S (2001) "Linkages Between Climate Change and Desertification in East Africa Part 2: Policy Linkages Between Climate Change and Desertification". *Arid Lands Newsletter,* No. 49, N.P Humid Savanna Zone International Development Research Centre.

Eriksen, S.H. (2005). "The Dynamics of Vulnerability: Locating Coping Strategies in Kenya and Tanzania". *Geographical Journal;* Vol. 171: 287-305.

Ewert F, et al (2002). "Effects of Elevated CO_2 and Drought on Wheat: Testing Crop Simulation Models for Different Experimental and Climatic Conditions". *Agriculture, Ecosystems and Environment;* Vol. 93: 249–266

Eriksen, S (2009) *Sustainable Adaptation: Emphasizing Local, Global Equity and Environmental Integrity,* Norway IHDP.

Fargione, J., Hill, J., Tilman, D., Polasky, S., Hawthorne, P., 2008. "Land clearing and the biofuel carbon debt". *Science* 319, 1235–1238.

FAO (2003) Responding to Agricultural and Food Insecurity Challenges Mobilising Africa to Implement Nepad Programmes. Conference of ministers of agriculture of the African Union- Maputo, Mozambique. 1-2 July, 2003.

FAO (2005) 31ˢᵗ Session of the Committee on World Food Security 23-26 May 2005 Special Event on *Impact of Climate Change, Pests and Diseases on Food Security and Poverty Reduction*; Rome-Italy.

FAO (2006) *Building Resilience for an Unpredictable Future: How Organic Agriculture Can Help Farmers Adapt To Climate Change*; Rome-Italy.

FAO. 2000a. Proceedings of the International Workshop on Community Forestry in Africa. *Participatory forest management: a strategy for sustainable forest management in Africa* 26-30 April 1999 Banjul, the Gambia. Rome.

FAO (2001). *State of Worlds Forest*: Rome Food and Agricultural Organization of the United Nations.

FAO (2007) *Adaptation to Climate Change in Agriculture, Forestry and Fisheries: Perspectives, Framework and Priorities:* Rome: FAO.

FAO (2008) *The State of Food and Agriculture. Biofuels: Prospects, Risks and Opportunities.* Rome: FAO.

FAO (2009) *Climate Change and Food Security in the Pacific.* Rome: FAO.

FAO (2009) *Food Security, and Agricultural Mitigation in Developing Countries: Options for Capturing Synergies* Rome: FAO.

Falkenmark, M. And Rockstro"m, J.(2004). *Balancing Water for Humans and Nature: The New Approach in Eco-hydrology.* London Earthscan Publications Ltd-UK.

Farm Africa. 2000. Phase II Kafa-Sheka Project Proposal. *Farm Ethiopia-Farm Africa.* Addis Ababa.

FBD (Forestry and Beekeeping Division, Ministry of Natural Resources & Environment. Government of Tanzania). 2001. *Guideline for Establishing Community Based Forest Management.* Dar es Salaam: Government printer.

Final Technical Report. Monte Alen Segmet, Equatorial Guinea, Monte Alen – Monts de Cristal Landscape (1) ad Maiko Tayna Kahuzi-Biega Landscape (10) Democratic Republic of Congo, Conservation International.

Fischer G et al (2002). Climate Change and Agricultural Vulnerability; Special Report as Contribution to the World Summit on Sustainable Development, Johannesburg 2002. Laxenburg:International Institute for Applied Systems Analysis.

Fischer, A et al (2005) "Will U.S. Agriculture Really Benefit from Global Warming? Accounting for Irrigation in the Hedonic Approach." *American Economic Review*; vol. 95 No.1: 395–406.

Gordon-Maclean, A., Laizer, J., Harrison, P.J. and Shemdoe, R. 2008. *Biofuel Industry Study, Tanzania.* World Wide Fund for Nature (WWF), Tanzania and Sweden.

Fuhrer, J (2006) "Agricultural Systems: Sensitivity to Climate Change", *CAB reviews: perspectives in Agriculture, Veterinary Science, Nutrition and Natural Resources;* Vol. 1. No. 052.

Furtado, J.I. et al (2000). *Economic development and environmental sustainability: policies and principles for a durable equilibrium.* Washington, D.C: World Bank, 116 pp.

Ghai, D., ed. (1994) "Development and environment: sustaining people and nature", *Development and Change,* 25 (1), 1–11.

Global Environmental Outlook, United Nations Environment Programme (UNEP), Oxford University Press, 1999.

Goldemberg, J., Squitieri, R., Stiglitz, J., Amano, A., Shaoxiong, X. and Saha, R. 1996, "Introduction: scope of the assessment", In *Climate Change 1995: Economic and Social Dimensions of Climate Change,* Contribution of Working Group III to the Second Assessment Report of the Intergovernmental Panel on Climate Change, Cambridge: Cambridge University Press.

Government of South Africa, 2003b. "Broad-Based Black Economic Act, 2003. *Government Gazette,* 1–5.

Government of South Africa, 2004. "National Environmental Management: Biodiversity Act No. 10 of 2004". *Government Gazette* 467, 1–85.

Gregory P.J et al (1999) "Managed Production Systems", In: Walker, B. et al 1999 (Eds). *The Terrestrial Biosphere and Global Change: Implications for Natural and Managed Systems.* Cambridge: Cambridge University Press:. 229–270.

Gregory P.J, et al (2005) Climate Change and Food Security. Philosophical Transactions of the Royal Society B: Biological Sciences; Vol. 360: 2139–2148.

Gregory, P.J et al (2009). "Integrating Pests and Pathogens into the Climate Change/ Food Security Debate", In Journal of Experimental Botany vol.60 No. 10: 2827-2838: London: Oxford University Press.

Grist, N (2008) "Positioning Climate Change in Sustainable Development Discourse", *Journal of International Development* Vol. 20: 783-803.

Grossman, D., Holder, P., 2006. Contract Parks in South Africa: Paper Prepared for IUCN Southern Africa Sustainable Use Specialist Group. 15 pp *Guinea, Monte Alen – Monts de Cristal Landscape (1) and Maiko Tayna Kahuzi.*

Grundmann, R (2007). "Climate Change and Knowledge Politics", *Environmental Politics* Vol. 16: 414–432.

Haile M. (2005) "Weather Patterns, Food Security and Humanitarian Response in Sub-Saharan Africa", *Philosophical Transactions of the Royal Society B: Biological Science;* Vol. 360: No.1463: 2169–2182.

Hahn M.B et al (2009) "The Livelihood Vulnerability Index: A Pragmatic Approach to Assessing Risks from Climate Variability and Change", *A Case Study in Mozambique: Global Environmental Change* Vol. 19: 74-88

Hall, A. (2008) "Paying for Environmental Services: The Case of Brazilian Amazonia", *Journal of International Development* Vol. 20: 965-981

Halsnaes, K and Traerup, S (2009) "Development and Climate Change: A Mainstreaming Approach for Assessing Economic, Social, and Environmental Impacts of Adaptation Measures." *Environmental Management Journal* Vol. 43: 765-778

Harmon, M.E., Ferrell, W.K., Franklin, J.F., 1990. "Effects on carbon storage of conversion of old-growth forests to young forests", *Science* 247, 699–702.

Hamza, M et al (2009) *Climate Change, Environmental Degradation and Migration:* Springer Science and Business Media

Hansen J. W (2005). Integrating Seasonal Climate Prediction and Agricultural Models for Insights into Agricultural Practice:, *Philosophical Transactions of the Royal Society B: Biological Sciences* Vol. 360: 2037–2047.

Hassan, R and C. Nhemachena, (2008) "Determinants of African farmers' strategies for adapting to climate change: multinomial choice analysis", *African Journal of Agricultural and Resource Economics* 2 (1) (2008) 83–104.

Hay R.K.M and J.R, Porter (2006) The Physiology of Crop Yield (2006) 2nd Edn. Oxford: Blackwell Publishing.

Hecht, S. and Cockburn, A. (1989) *The* Fate of the Forest: *Developers, Destroyers and Defenders of the Amazon*, London: Verso.

Hepburn, C and Stern, N (2008) "A New Global Deal on Climate Change", *Oxford Review of Economic Policy* Vol. 24 No. 2:259-279.

Hesse, C. & P. Trench. 2000a. Decentralisation and institutional survival of the fittest in the Sahel – what hope CPRM? Paper presented to the Eighth Biennial Conference of the International Association for the Study of Common Property, Bloomington, Indiana, May 29-June 4 2000.

Hoppers, C.A.O., 2007. Intellectual property, social justice and global development: implications for policy and educators in the north and the south. Address to the Finlandssvensk Utbildningsconferens on Knowledge, Competence and Cooperation, 19–20 April 2007, Annaholmen Conference Centre. Helsinki. 13 pp.

Houghton, J.T et al., (2001) J.T. Climate Change: The Scientific Basis, Cambridge University Press, Cambridge

Hulme, M., R. et al (2001) 'African Climate Change: 1900-2100', Climate Research; Vol. 17, No. 2: 145-68

Hulme D, Murphree M (Eds) (2001) African Wildlife and Livelihoods: The Promise and Performance of Community Conservation. Heinemann, Portsmouth.

Huq, S., A. Rahman, M. Konate, Y. Sokona and H. Reid (2003). *Mainstreaming Adaptation to Climate Change in Least Developed Countries* (LDCs). IIED, London.

Huq S et al (2006) Climate Change and Development Links. Gatekeeper Series No. 123. International Institute for Environment and Development (IIED), London, UK, p 24.

Hyden, G. and Mugabe, J. 'Governance and Sustainable Development in Africa: The Search for Economic and Political Renewal in Okoth-Ogendo," H.W.O. and Tumushabe, G. W. eds. Governing the Environment: Political Change and Natural Resources Management in Eastern and Southern Africa Nairobi: African Centre for Technological Studies, 1999.

Iddi, S. 2000. Community involvement in forest management: first experiences from Tanzania. The Gologolo Joint Forest Management Project: A case study from the West Usambaras Mountains. In FAO 2000a.

IIED, (1994) Whose Eden: An overview of Community Approaches to Wildlife Management International Institute for Environment and Development (IIED).UK, 38.

International Institute for Environment and Development (IIED) (2009) *Community management of natural resources in Africa: Impacts, Experiences and Future Directions.* London IIED UK.

Ingram, J.S.I. et al (2008) The Role of Agronomic Research in Climate Change and Food Security Policy. Agriculture, Ecosystems and Environment; Vol.126: 4–12.

International Institute for Environment and Development (IIED) (2005). Adapting to Climate Change in East Africa: A Strategic Approach. Gate Keepers Series No. 117: IIED-UK.

International Institute for Environment and Development (IIED) (2006) *Biodiversity, Climate Change and Complexity: An opportunity for Securing Co-Benefits: Sustainable Development Opinion.* NP: NP.

International Institute for Environment and Development (IIED) (2007) Migration and Adaptation to Climate Change: Sustainable Development Opinion. IIED-UK.

International Institute for Environment and Development (IIED) (2007) *Climate Change and Adaptation in the Niger River Basin*: IIED-London.

International Institute for Environment and Development (IIED) (2004) *Climate Change, Biodiversity and Livelihood Impacts*: IIED-London.

International Institute for Environment and Development (IIED) and World Wide Fund for Nature(WWF) (2007) *Climate, Carbon, Conservation and Communities. London:* IIED/WWF-London.

Intergovernmental Panel on Climate Change (IPCC), 2007a. *Fourth Assessment Report. Climate Change 2007: Synthesis Report.* Geneva: IPCC.

Intergovernmental Panel on Climate Change (IPCC). (2002). *Climate change and biodiversity*, ed. H. Gitay, A. Suárez, R. T. Watson & D. J. Dokken. IPCC Technical Paper V. Geneva, Switzerland & Nairobi, Kenya, World Meteorological Organization (WMO) & United Nations Environment Programme (UNEP). IPCC. (2007). *Climate change 2007 – impacts, adaptation and vulnerability.* Contribution of Working Group II to the Fourth Assessment Report of the IPCC. Cambridge, UK, Cambridge University Press.

International Institute for Sustainable Development (IISD) (2003), *International Union for the Conservation of Nature and Natural Resources, Stockholm Environment Institute. Livelihoods and climate change.* Manitoba IISD.

IPCC (2001) Climate Change 2001: Impacts, Adaptation, and Vulnerability, Cambridge: Cambridge University Press.

IPCC Working Group I. (2007) Climate Change 2007: The Physical Science Basis, Summary for Policymakers.

Klein, J.T et al (2005) Integrating Mitigation and Adaptation into Climate and Development Policy: Three Research Questions. Environmental Science & Policy Vol 8, No. 6, 579-588. Elsevier.

International Institute for Applied Systems Analysis (IIASA) (2009) *Biofuels and Food Security: Implications of an Accelerated Biofuel Production.* Vienna:IIASA.

International Institute for Environment and Development (IIED) (2009) Community Management of Natural Resources in Africa: Impacts, Experiences and Future Directions: IIED; United Kingdom.

IUCN, 2007. "Forests and Livelihoods: Reducing Emissions from Deforestation and Ecosystem Degradation (REDD)", *Climate Change Briefing.* World Conservation Union.

Irmak A, et al (2005). "Evaluation of the CROPGRO-Soybean Model for Assessing Climate Impacts on Regional Soybean Yields", *Transactions of the ASAE;* Vol. 48: 2343–2353.

Jagadish S.V.K, et al (2008) "Phenotyping Parents of Mapping Populations of Rice for Heat Tolerance During Anthesis", *Crop Science;* Vol. 48: 1140–1146.

Jagtap SS, and J.W, Jones (2002). "Adaptation and Evaluation of the CROPGRO-Soybean Model to Predict Regional Yield and Production", *Agriculture Ecosystems and Environment;* Vol. 93: 73–85.

Jewitt, S., 1995. "Voluntary and official forest protection committees in Bihar: solutions to India's deforestation?" *Journal of Biogeography* 22 (6), 1003–1021.

Jiggins, J. (1989). "An examination of the impact of colonialism in establishing negative values and attitudes towards indigenous agricultural knowledge", In Warren, D. M., Slikkerveer, L. J., and Titilola, S. O. (eds.), *Indigenous Knowledge Systems: Implications for Agriculture and International Development,* Technical and Social Change Program, Iowa: Iowa State University Reserve Fund, 102–140.

Johnson SN, et al (2008). "Varietal Susceptibility of Potatoes to Wireworm Herbivory", *Agricultural and Forest Entomology;* Vol. 10: 167–174.

Jones B. 2004b. *Synthesis of the current status of CBNRM Policy and Legislation in Botswana, Malawi, Mozambique, Namibia, Zambia and Zimbabwe,* WWFSARPO.

Jones P.G, Thornton P.K (2003). "The Potential Impacts of Climate Change on Maize Production in Africa and Latin America in 2055", *Global Environmental Change;* Vol. 13: 51–59.

Jones, B., and M. Murphree (2001) The Evolution of Policy on Community Conservation in Namibia and Zimbabwe. In D. Hulme, and M. Murphree, editors.

Jones, P.G and P.K Thornton, (2003) "The Potential Impacts of Climate Change in Tropical Agriculture: The Case of Maize in Africa and Latin America in 2055", *Global Environmental Change* Vol. 13: 51–59.

Kahn, M.E., (2003) "Two Measures of Progress in Adapting to Climate Change", *Global Environmental Change;* Vol. 13: 307-312.

Kaiser, H. M, et al (1993a). «Adaptation to Global Climate Change at the Farm Level", In Kaiser, H.M and T. Drennen, (Eds) 1993. *Agricultural Dimensions of Global Climate Change.* Delray Beach, Fl.: St. Lucie Press.

Kaiser, H. M, et al (1993b). A Farm-Level Analysis of Economic and Agronomic Impacts of Gradual Warming. *American Journal of Agricultural Economics;* Vol. 75 No. 2: 387–98.

Kajembe, G.C., Nduwamungu, J., Luoga, E.J., 2005. "The impact of community-based forest management and joint forest management on the forest resource base and local people's livelihoods: case studies from Tanzania", Commons Southern Africa occasional paper series No 8. Centre for Applied Social Sciences, University of Zimbabwe/Programme for Land and Agrarian Studies, University of the Western Cape, Harare/Cape Town. 22 pp.

Kamanga, K.C. 2008. *The Agrofuel Industry in Tanzania: A Critical Enquiry into Challenges and Opportunities.* A research report. Hakiardhi and Oxfam Livelihoods Initiative for Tanzania (JOLIT), Dar es Salaam.

Kass, N.E., 2001. "An ethics framework for public health", *American Journal of Public Health* 91, 1776–1782.

Katerere, Y., E. Guveya & K. Muir. 1999. Community forest management: Lessons from Zimbabwe. Issue paper No. 89 Drylands Programme IIEDE.

Keeney, D and Nanninga C (2008) *Biofuel and Global Biodiversity.* Minnesota Institute for Agriculture and Trade Policy,

Kemanian AR, et al (2005). "Transpiration-Use Effciency of Barley", *Agricultural and Forest Meteorology;* Vol. 130: 1–11.

Klein, R.J.T. and J.B. Smith, (2003) "Enhancing the Capacity of Developing Countries to Adapt to Climate Change: A Policy Relevant Research Agenda", In Smith, J.B et al (Eds) (2003) *Climate Change, Adaptive Capacity and Development,* London: Imperial College Press, 317 334.

Koh, L.P., Ghazoul, J.,(2008). "Biofuels, Biodiversity, and People: Understanding the conflicts and finding opportunities", *Biological Conservation.* 141, 2450–2460.

Kothari, A., Pathak, N., Anuradha, R. V. and Taneja, B. (1998) *Communities and Conservation: Natural Resource Management in South and Central Asia,* New Delhi: Sage Publications,

Kurukulasuriya, P and R. Mendelsohn (2008). "A Ricardian Analysis of the Impact of Climate Change on African Cropland", *African Journal of Agricultural and Resource Economics,* Vol. 2, No.1: 1–23.

Kurukulasuriya, P. et al (2006). "Will African Agriculture Survive Climate Change?" *The World Bank Economic Review:* Vol. 20 No.3: 367-388. London-Oxford University Press.

Kurukulasuriya, P., and S. Rosenthal. 2003a. "Climate Change and Agriculture: A Review of Impacts and Adaptations." Climate Change Series 91. Environment Department Papers, World Bank, Washington, D.C.

Larsen, K and U.G-Ostling (2009) "Climate Change Scenarios and Citizen participation: Mitigation and Adaptation Perspectives in Constructing Sustainable Futures", *Habitat International Journal* 333, 260-266; Sweden, Elsevier.

Larsson, T.B, A. Barbati, J. Bauhus, J. Van Brusselen, M. Lindner, M. Marchetti, B. Petriccione, and H. Petersson (2007) "Role of Forests in Carbon Cycles, Sequestration and Storage: Climate Change Mitigation, Forest Management and Effects on Biological Diversity", *Newsletter No.5;* Canada.

Leach, M., Mearns, R. and Scoones, I. (1999) "Environmental entitlements: dynamics and institutions in community-based natural resource management", *World Development,* 27 (2), 225–247.

Lele, S. (2000) Godsend, *Sleight of Hand, or Just muddling Through: Joint Water* and Forest *Management in India Natural Resource Perspectives.* 53, London: ODI, .

Lemenih, M and M. Bekele (2008). *Participatory Forest Management: Best Practices, Lesson Learnt And Challenges Encountered: The Ethiopian and Tanzanian Experiences:* FARM-Africa/SOS-Sahel.

Lempriere, T.C; P.Y Bernier; A.L Carroll; M.D Flannigan; E.H Hogg; R.P Gilsenan; D.W McKenney; J.H Pedlar and D.Blain (2008) *The Importance of Forest Sector in Adapting to Climate Change: Canadian Forest Service:* Canada.

Li, T. M. (1996) "Images of community: discourse and strategy in property relations, *Development and Change",* 27 (3), 501–528. International CBNRM Workshop, Washington, DC, 10–14 May 1998.

Li, T. M. (2002) "Engaging simplifications: community-based resource management, market processes and state agendas in upland Southeast Asia", *World Development,* 30 (2), 265–283.

Ligon, E., Narain, U., 1999. "Government management of village commons: comparing two forest policies", *Journal of Environmental Economics and Management* 37 (3), 272– 289.

Lobell, D and M.B. Burke, (2008) "Why are Agricultural Impacts of Climate Change So Uncertain? The Importance of Temperature Relative to Precipitation", *Environmental Research Letters,* Vol. 3 No. 3, Stanford, USA.

Lobell, D. et al, (2009) "Climate Extremes and Crop Adaptation: Summary of the Statement From a Meeting at the Programme on Food Security and Environment", Stanford, CA. June, 16-18, 2009.

Long SP et al (2004) "Rising Atmospheric Carbon Dioxide: Plants Face the Future", *Annual Review of Plant Biology;* Vol. 55: 591–628.

Lubell, M., 2002. "Environmental activism as collective action", *Environment and Behaviour* 34 (4), 431– 454.

Ludwig, D., Hilborn, R., Walterns, C., (1993). "Uncertainty, resource exploitation, and conservation: Lessons from History", *Science Vol.* 260 No. 17: 36.

Lundy, P., 1999. "Community participation in Jamaican conservation projects", *Community Development Journal* 35, 122–132.

Luo QY, Lin E. "Agricultural vulnerability and adaptation in developing countries: the Asia-Pacifc region", *Climatic Change* (1999) 43: 729–743.

Lumley S, Armstrong P. (2004). "Some of the Nineteenth Century Origins of the Sustainability Concept", *Environment, Development and Sustainability* Vol. 6 No. 3: 367–378.

MA. 2005. *Millennium Ecosystem Assessment: Ecosystems and Human Well-being,* Synthesis, Island Press.

Maarten et al (2008) "Community level adaptation to climate change", *The Potential Role of Participatory Community Risk Assessment:* Vol. 18: 165-179.

Makurira, H. et al (2007b). "Towards a Better Understanding of Water Partitioning Processes for Improved Smallholder Rainfed Agricultural Systems: A Case Study of Makanya Catchment, Tanzania", *Physics and Chemistry of the Earth* Vol. 32, No. 15–18: 1082–1089.

Mansourian, S; A. Belokurov and P.J Stephenson (2009) *The Role of Forest Protected Areas in Adapting to Climate Change.* Food and Agriculture Organisation.

Martin, R. B. (1986) 'Communal Areas Management Programme for Indigenous Resources (CAMPFIRE)'. Working Document 1/86. Harare: Department of National Parks and Wild- life Management.

Massawe, E. 2000. *Mgori Forest: The Current Situation and its Future After the Donors have* Left. Bogor, Indonesia.

Mathews, E. 2001. Understanding the FRA 2000. World Resources Institute Forest Briefing No. 1.

Matsui T, et al (1997) "High Temperature-Induced Spikelet Sterility of Japonica Rice at Flowering in Relation to Air Temperature, Humidity and Wind Velocity Conditions", *Japanese Journal of Crop Science* (1997) 66: 449–455.

Matta J.R and RRA, Janaki (2006) "Perceptions of Collective Action and its Success in Community Based Natural Resource Management: An empirical analysis", Elsevier: *Forest Policy and Economics,* Vol. 9: 274– 284.

Mauambeta, D. 2000. "Sustainable management of indigenous forests in Mwanza East, Malawi", *In FAO 2000a.*

Mayers, J., Evans, J. & T. Foy. 2001. *Raising the stakes Impacts of Privatisation, certification and partnerships in South African forestry* London: IIED,

McGuire, S.J (2007) "Vulnerability in Farmer Seed Systems: Farmer Practices for Coping with Seed Insecurity for Sorghum in Eastern Ethiopia", *Economic Botany* Vol. 6: 211–222.

McGuire, S.J (2008) "Securing Access to Seed: Social Relations and Sorghum Seed Exchange in Eastern Ethiopia", *Human Ecology.* 36, 217–229.

McGuire, S.J and Sperling, L (2008) "Leveraging Farmers' Strategies for Coping with Stress: Seed Aid in Ethiopia", *Global Environmental Change* Vol.18, No: 4 679-688.

McNeely, J. (1995) "IUCN and indigenous peoples: how to promote sustainable development", in D. M. Warren, L. J. Slikkerveer and D. Brokensha, eds, *The Cultural Dimension of Development: Indigenous Knowledge Systems,* London: Intermediate Technology Publications.

Mearns L.O (Ed) 2003. Issues in the Impacts of Climate Variability and Change on Agriculture. Applications to the South-Eastern United States; Dordrecht-Kluwer Academic Publishers.

Mehlman P, Kernan C, Bonilla JC. 2006. *Conservation International CARPE USAID.*

Mendelsohn, R and D.A, Dinar (1999) "Climate Change, Agriculture, and Developing Countries: Does Adaptation Matter?" *The World Bank Research Observer,* Vol.14, No. 2: 277-293. USA-The World Bank.

Metcalfe, S (1993) "CAMPFIRE: Zimbabwe's Communal Areas Management Programme For Indigenous Resources", a paper prepared for the Liz Claiborne & Art Ortenberg Foundation workshop on Community-Based Conservation, July 1993, Zimbabwe.

MINEF (Ministry of Environment and Forests). 1998. *Manual of the Procedures for the Attribution, Norms and Management of Community Forests.* Government of Cameroon.

Montgomery S., M. Lucotte and I. Rheault (2000). "Temporal and spatial influences of flooding on dissolved mercury in boreal reservoirs", *The Science of the Total Environment* Vol. 260 No. 1-3: 147-157.

Morris, P.M., 2002. "The capabilities perspective: a framework for social justice", *Families in Society* 83, 365–373.

Morton, J.F. (2007) "The Impact of Climate Change on Smallholder and Subsistence Agriculture"", Proceedings of the National Academy of Sciences of the United States of America Vol. 104: No. 50: 19680-5.

Mortreux, C and Barnett, J (2009) "Climate Change, Migration and Adaptation in Funafuti, Tuvalu", *Global Environmental Change* Vol. 19: 105-112.

Mwandosya, M.J. et al (1998) The *Assessment of Vulnerability and Adaptation to Climate Change Impacts in Tanzania NP: Dat es Salaam*

Nana, A. 2000. An example of cooperation between government and non-government institutions in carrying out community forest management activities. The Case of Naturama's activities in the Kabore Tambi National Park in Bukina Faso. In FAO 2000a.

Naughton-Treves, L., et al (2006). *Expanding protected areas and incorporating human resource use: a study of 15 forest parks in Ecuador and Peru. Sustainability:*

Nelson, V and T. Stathers (2009) "Resilience, Power, Culture and Climate: A case Study From Semi-Arid Tanzania, and New Research Directions", *Gender and Development Journal* Vol. 17 No. 4 81-94.

Neumann R. 1998. *Imposing Wilderness: Struggles Over Livelihood and Nature Preservation in Africa*, N:P University of California Press.

Newton AC et al (2009). Deployment of Diversity for Enhanced Crop Function. Annals of Applied Biology (2009) (in Press).

Nhemachena, C. And R, Hassan (2007). Micro-Level Analysis of Farmers' Adaptation to Climate Change in Southern Africa; IFPRI Discussion Paper No. 00714-Washington, DC : International Food Policy Research Institute,

Ngigi, S.N. (2003). "What is the Limit of Up-scaling Rainwater Harvesting in a River Basin?" *Physics and Chemistry of the Earth* Vol. 28, No. 20–27: 943–956.

Nhemachena, C. And R, Hassan (2007). Micro-Level Analysis of Farmers' Adaptation to Climate Change in Southern Africa; IFPRI Discussion Paper No. 00714-International Food Policy Research Institute, Washington, DC.

Ntsime, P.T (2004). "Deconstructing Sustainable Development: Towards a Participatory Methodology for Natural Resource Management. South Africa", *Informa World Journal* Vol. 21: No. 4: 717-718.

Nygren, A., 2000. "Development discourses and peasant–forest relationships: natural resource utilization as social process", In: Doornbos, M., Saith, A., White, B. (Eds.), Forests: Nature, People, Power. London: Blackwell Publishers, 11–33.

O'Brien, K. et al (2000). Is Information Enough? User Responses to Seasonal Climate Forecasts in Southern Africa. Oslo: Centre for International Climate and Environmental Research (CICERO), University of Oslo, Report 2003: 3.

Ogungo, P.O (2007) *Participatory Forest Management in Kenya: Is There Anything for the Poor? Proceedings:* International Conference on Poverty Reduction and Forests, Bangkok, September 2007.

Olsen, K., H. Ekwoge, R. Ongie, J. Acworth, E. O'Kah & C. Tako. 2001. A Community Wildlife Management Model from Mount Cameroon. ODI RDFN Paper. No. 25e.

Olsen, S.B. et al (2006). *Ecosystem-Based Management: Markers for Assessing Progress.* The Hague: UNEP/GPA.

Orindi, V.A and Murray, L.A (2005) Adapting to Climate Change in East Africa: A Strategic Approach. GateKeepers Series No. 117: International Institute for Environment and Development.

Ormerod S.J and I. Durance (2009). Restoration and Recovery from Acidification in Upland Welsh Streams over 25 years: Journal of Applied Ecology Vol. 46: 164–174.

Ormerod, S. J (2009) (Ed) Climate Change, River Conservation and the Adaptation Challenge: Aquatic Conservation: Marine and Fresh Water Ecosystems Vol. 19: 609-613.

Oxfam International 2008. *Another Inconvenient Truth. Available at: http:// www.oxfam.org.uk/resources/policy/climate_change/downloads/* bp114_inconvenient_truth.pdf.

Oxfam. 2008. *Viet Nam: "Climate Change, Adaptation and Poor People",* Hanoi and Oxford: Oxfam. 56 pp: *In Land and Water Resource Management in Asia: Challenges for Climate Change.* International Institute for Sustainable Development (Tyler, S and L. Fajbar 2009).

Pandey, D.N.(2002). "Carbon Sequestration in Agro Forestry System", *Climate Policy* Vol: 7 1–12.

Parker W.E, and J.J Howard (2001). "The Biology and Management of Wireworms (Agriotes spp.) on Potato with Particular Reference to the UK", *Agricultural and Forest Entomology;* Vol. 3: 85–98.

Parnell R. 2006. *Mayumba National Park CARPE Agreement Final Technical Report: October 1, 2003 – September* 30, 2006, WCS / USAID.

Parry, J; Hammill, A and Drexhage, J (2005) *Climate Change and Adaptation.* Manitoba: International Institute for Sustainable Development.

Parry M.L (2004). "Effects of Climate Change on Global Food Production under SRES emissions and socio-economic scenarios", *Global Environmental Change,* Vol.14: 53–67.

Parry, M.L et al (1999) "Climate Change and World Food Security: a New Assessment", *Global Environmental Change;* Vol. 9 No. 1: 51-6

Pascual M., et al. Malaria Resurgences in the East Africa Highlands: Temperature Trends Revisited. Proceedings of the National Academy of Sciences Vol. 103: 5829–34.

Patt A.G, et al (2009). Effects of Seasonal Climate Forecasts and Participatory Workshops among Subsistence Farmers in Zimbabwe. Proceedings of the National Academy of Sciences, USA, Vol. 102: 12623–12628

Peck, J (1999) "Measuring Justice for Nature: Issues in Evaluating and Litigating Natural Resources Damages", *Journal of Land Use & Environmental Law Quarterly* Vol. 1: 35-45.

Peters, P. 1986 *Failed Magic or Social Context? Market Liberalization and the rural poor* in Malawi Mimeo.

Phillips, J and B.McIntyre (2000) "ENSO and Inter-Annual Rainfall Variability in Uganda: Implications for Agricultural Management", *International Journal of Climatology;* Vol. 20: 171-182, USA-Columbia University.

Pierre, J. and Peters, B. Guy. *Governance, Politics and the State.* London: Macmillan, 2000.

Porter JR, and M.A, Semenov (2005). "Crop Responses to Climatic Variation", *Philosophical Transactions of the Royal Society;* Vol. 360: 2021–2035.

Pretty, J., 2003. *Social Capital and the Collective Management of Resources.* Science 302, 1912– 1914.

Rabetaliana H. & P. Schachenmann. 2000. "Community-Based Management of Natural and Cultural Resources in Ambondrombe – A Historic Site in Madagascar", In *CM News No. 4 Gland*

Ravnborg, H.M., Westermann, O., (2002). "Understanding interdependencies: stakeholder identification and negotiation as a precondition to collective natural resource management", *Agricultural Systems* 73, 41–56.

Raymond J. Kopp & V. Kerry Smith (1993), "Understanding Damages to Natural Assets", *in Valuing Natural Assets, the Economics of Natural Resource Damage Assessment* Vol. 6: Pgs 10-11.

Reid, H (2004) *Climate Change, Biodiversity and Livelihood Impacts.* International Institute for Environment and Development.

Reid, H., B. Pisupati and H. Baulch (2004). "How Biodiversity and Climate Change Interact", *SciDev.Net Biodiversity Dossier Policy Brief.*

Reilly, J., et al. (1995), "Agriculture in a Changing Climate: Impacts and Adaptation", *Climate Change*, Washington NP.

Ribot, J.C., 1999. "Accountable representation and power in participatory and decentralized environmental management", *Unasylva* 50, 18–22.

Risbey, J. S (2008) "The New Climate Discourse: Alarmist or alarming?" *Global Environmental Change* Vol.18: 26-37.

Robbins P. et al (2006) Even Conservation Rules are Made to be Broken: Implications for Biodiversity. Environmental Management Journal: Vol. 37, No. 2: 162–169.

Robinson, E.J.Z and F. Maganga (2009). "The implications of improved", communications for participatory forest management in Tanzania: Blackwell Publishing Ltd, African. *Journal of Ecology;* Vol. 47: Suppl. 1: 171–178.

Roe, D (2006) "Biodiversity, Climate Change and Complexity: An Opportunity For Securing Co-Benefits", International Institute for Environment and Development.

Roe, D ; H. Reid; K. Vaughan; E. Brickell; J. Elliott (2007) Climate, Carbon, Conservation and Communities. IIED/WWF Briefing.

Saldana-Zorrilla (2008) "Stakeholders' Views in Reducing Rural Vulnerability to Natural Disasters in Southern Mexico: Hazard Exposure, Coping and Adaptive capacity", *Global Environmental Change* Vol. 18 No.4: 583-597

Salinger, J, et al (Eds.) (2005) "Increasing Climate Variability and Change: Reducing the Vulnerability of Agriculture and Forestry", Reprinted from *Climatic Change,* Vol. 70, Nos. 1-2, VI: 362.

Sanoff, H., 2000. *Community Participation Methods in Design and Planning.* N.P: N.P.

Sathaye, J., A. Najam, C. Cocklin, T. Heller, F. Lecocq, J. Llanes-Regueiro, J. Pan, G. Petschel-Held , S. Rayner, J. Robinson, R. Schaeffer, Y. Sokona, R. Swart, H. Winkler, 2007: "Sustainable Development and Mitigation", In *Climate Change 2007: Mitigation. Contribution of Working Group III to the Fourth Assessment Report of the Intergovernmental Panel on Climate Change* [B. Metz, O.R. Davidson, P.R. Bosch, R. Dave, L.A. Meyer (eds)], Cambridge University Press, Cambridge, United Kingdom and New York, NY, USA.

Sathaye, Jayant; Shukla P.R and Ravindranath, N. H (2006) "Climate Change, Sustainable Development and India", Global and National Concern Vol. 90 No.3.

Sayer, J. and Campbell, B. 2004. *The Science of Sustainable Development: Local Livelihoods and the Global Environment.* Cambridge: Cambridge University Press.

Scherm H, et al (2000) "Global Networking for Assessment of Impacts of Global Change on Plant Pests" *Environmental Pollution;* Vol. 108: 333–341.

Science, Practice & Policy 2, 32–44.

Scoones, I. 1998. Sustainable Rural Livelihoods: A Framework for Analysis. Institute of Development Studies, IDS Working Paper No. 72.

SEI, 2005. *Sustainable Pathways to Attain the Millennium Development Goals: Assessing the Key Role* of Water, Energy and Sanitation. Stockholm Stockholm: Environ- ment Institute.

Siegenthaler U, et al (2005). "Stable Carbon Cycle–Climate Relationship During the Late Pleistocene Science", Vol. 310: 1313–1317.

Slater R et al (2007). "Climate Change, Agricultural Policy and Poverty Reduction - How Much Do We Know?" *Natural Resource Perspectives* Vol.109: 1–6.

Slingo, A.J. et al (2005) "Introduction: Food Crops in a Changing Climate", *Philosophical Transactions of the Royal Society, Series* B; Vol. 360: 1983–1989.

Singh, P.P (2008) "Exploring Biodiversity and Climate Change Benefits of Community-Based Forest Management", *Global Environmental Change Journal* Vol. 18:468-478.

Siry, J.P; R. De La Torre; R.L Izlar (2009) The Role of Managed Forests in Climate Change Mitigation. XIII World Forestry Congress.

Smit B. and O. Pilifosova. (2001). "Adaptation to Climate change in the context of sustainable development and *equity", In Climate change* 2001: *impacts, adaptation, and vulnerability.* J.J. McCarthy, O.F. Canziani, N.A. Leary, D.J. Dokken, and K.S. White (editors). Intergovernmental Panel on Climate Change, Cambridge University Press, New York, N.Y. 876–912.

Smit, B., Burton, I., Klein, R., Wandel, J., 2000. "An anatomy of adaptation to climate change and variability. Climatic Change 45, 223–251.

Smit, B., I. Burton, R.J.T. Klien, and J. Wandel. 2000. An anatomy of adaptation to climate change and variability", *Climatic Change* 45: 223–251.

Smit, B et al., (1996) "Agricultural Adaptation to Climatic Variation", *Climatic Change;* Vol. 33: 7–29.

Smith, D. and J. Troni (2004). *Climate Change and Poverty: Making Development Resilient to Climate Change.* London: DFID.

Smith, L. et al (2003). "Applying a social justice framework to college counseling center practice", *Journal of College Counseling* 6, 3–13.

Songorwa, A.N., 1999. "Community-based wildlife management (CWM) in Tanzania: are the communities interested?" *World Development* 27 (12), 2061–2079.

Sonko, K. and K. Camara. 1999. *Community Forestry Implementation in The Gambia: Its Principles and Prospects* in FAO 2000.

Soussana J.F, and U.A Hartwig (1996). "The Effects of Elevated CO_2 on Symbiotic N_2 Fixation: A Link Between the Carbon and Nitrogen Cycles in Grassland Ecosystems", *Plant and Soil* Vol. 187: 321–332.

Southern Africa Development Community (SADC), 1999. *Protocol on Wildlife Conservation and Law Enforcement.* SADC, Maputo. 14 pp.

Southern Africa Development Community (SADC), 2002. *Southern African Development Community Protocol on Forestry.* SADC, Luanda. 14 pp.

Spahni R et al (2005). "Atmospheric Methane and Nitrous Oxide of the Late Pleistocene from Antarctic Ice Cores", *Science*; Vol. 310: 1317–1321

Sperling, H.D et al (2008). "Moving Towards More Effective Seed Aid", *Journal of Development Studies* Vol. 44: 586–612.

Sperling, F. (2003). *Poverty and Climate Change: Reducing the Vulnerability of the Poor Through Adaptation.* Washington DC.: World Bank.

Staley JT et al (2007) "Effects of Summer Rainfall Manipulations on the Abundance and Vertical Distribution of Herbivorous Soil Macro-Invertebrates", *European Journal of Soil Biology* ; Vol. 43: 189–198.

Steel L. 2008. *Salonga-Lukenie-Sankuru Landscape. Luilaka River CBNRM Zone: Strategy Document for the Development of a Co-Management Plan,* USAID / CARPE / ICCN / WWF / WCS / La Societe Zoologique de Milwaukee / PACT.

Stern, N. (2006) "Supporting and Financing Africa's Resurgence", *Journal of African Economies* Vol. 15, No. 2: 161. London, Oxford University Press.

Stern, N. (2006). The Stern Review on the Economic Effects of Climate Change (Report to the British Government). Cambridge: *Cambridge University Press,*

Stern. N (2007). The Economics of Climate Change: The Stern Review. Cambridge University Press: Cambridge, UK.

Stone RC, and H. Meinke (2005). "Operational Seasonal Forecasting of Crop Performance", *Philosophical Transactions of the Royal Society* B: Biological Sciences; Vol. 360: 2109–2124.

Susskind, L., Cruikshank, J., 1987. *Breaking the impasse: consensual approaches to resolving public* disputes. New York: Basic Books, 276 pp.

Sutherst R et al (2007). "Pests under Global Change: Meeting Your Future Landlords?" In Canadell, J.G (Ed) 2007. *Terrestrial Ecosystems in a Changing World.* Berlin: Springer. 211–225.

Stuart, M.D and P.M Costa (1998) *Climate Change Mitigation by Forestry: A Review of International Initiative.* International Institute for Environment and Development.

Sulle, E and Nelson, F (2009) *Biofuels, Land Access and Rural Livelihoods in Tanzania:* London. International Institute for Environment and Development.

Taylor R, Murphree MW. 2007. *Case studies on successful southern African NRM initiatives and their impact on poverty and governance: Masoka and Gairesi case studies Zimbabwe,* IUCN / USAID FRAME.

Thomas, D.S.G and C, Twyman, (2005) "Equity and Justice in Climate Change Adaptation amongst Natural-Resource-Dependent Societies", *Global Environmental Change;* Vol. 15, No. 2: 115–124.

Thompson, J. (1995) "Participatory approaches in government bureaucracies: facilitating the process of institutional change", *World Development,* 23 (9), 1521–1554.

Thornton, P.K et al (2006) "Mapping Climate Vulnerability and Poverty in Africa; Report to the Department for International Development", *ILRI,* Nairobi, Kenya, May 2006, 200 pp.

Thornton, P.K et al (2009) "Spatial Variation of Crop Yield Response to Climate Change in Eastern Africa", *Global Environmental Change* Vol.19 No.1: 54-65. Sweden-Elsevier.

Tubiello FN et al (2007b). "Crop and Pasture Response to Climate Change", *Proceedings of the National Academy of Sciences,* USA, Vol. 104: 19686–19690.

Tubiello, FN, and F. Ewert (2002). "Simulating the Effects of Elevated CO_2 on Crops: Approaches and Applications for Climate Change", *European Journal of Agronomy;* Vol. 18: 57–74.

Tumushabe, G.W. and Okoth-Ogendo, H.W.O. "Governing the Environment: Political Change and Natural Resources Management in Eastern and Southern Africa." African Centre for Technological Studies, Nairobi, Kenya, 1999.

Tschakert, P et al (2009) "Holistic, Adaptive Management of the Terrestrial Carbon Cycle at Local and Regional Scales", *Global Environmental Change:* Vol. 18: 128-141

Tyler, S and L. Fajbar (2009) *Land and Water Resource Management in Asia: Challenges for Climate Change.* International Institute for Sustainable Development.

UFD (Uganda Forest Department). 2000. Collaborative Forest Management Agreement between the Forest Department and Bumusili Village regarding the management of Namatale Forest Reserve, Kampala.

United Nations. 2007. *The Millennium Development Goals Report 2007. New York, USA.*

United Nations Development Programme (UNDP),1997. *UNDP guidebook on participation.* 8 pp.

United Nations, 1992. Agenda 21 - "The United Nations Programme of Action from Rio", *The United Nations,* New York. 294 pp.

United Nations, 2006. *Social Justice in an Open World: the Role of United Nations.* The United Nations, New York. 146 pp.

United Republic of Tanzania (2006) *Participatory forest management in Tanzania: Facts and figures.* Produced by the Extension and Publicity Unit, Forest and Bee Keeping Division, Ministry of Natural Resources and Tourism, July.

Uphoff, N. (1998) 'Community-based natural resource management: connecting micro and macro processes, and people with their environments', Plenary Presentation.

URT (2007) *National Adaptation Programme of Action* (NAPA), Dar es Salaam. Vice-President Office-Tanzania Printing House.

Unruh, J.D (2008) "Carbon Sequestration in Africa: The Land Tenure Problem", *Global Environmental Change* Vol. 18: 700-707

Unruh, J.D. et al (2005). "Migrant Land Rights Reception and 'Clearing to Claim' in Sub-Saharan Africa: A Deforestation Example from Southern Zambia:, *Natural Resources Forum* Vol. 29,: 190–198. Wall, E and Smit, B (2005) Climate Change Adaptation in Light of Sustainable Agriculture. Journal of Sustainable Agriculture: Vol. 27 No. 1: The Haworth Press, Inc.

Vara Prasad PV et al (2000). "Effects of Short Episodes of Heat Stress on Flower Production and Fruit-set of Groundnut *(Arachis hypogaea L.)*", *Journal of Experimental Botany*; Vol. 51: 777–784.

Venter, O., Meijaard, E., Possingham, H., Dennis, R., Sheil, D., Wich, S., Hovani, L., Wilson, K., 2009. Carbon payments as a safeguard for threatened tropical mammals. Conservation Letters.

Verdin J, et al (2005) "Climate Science and Famine Early Warning. Philosophical Transactions of the Royal Society", B: *Biological Sciences;* Vol. 360: 2155–2168.

Vogel C and J. Smith (2002). "The Politics of Scarcity: Conceptualising the Current Food Security Crisis in Southern Africa", *South African Journal of Science;* Vol. 98: 315–317.

Vogt, G. & Vogt. K. 2000. *Hannu Biyu Ke Tchuda Juna – Strength in Unity.* Shared management of common property resources. A Case Study from Takieta, Niger. Securing the Commons No. 2 SOS Sahel Programme. Washington.

Von Braun J. (2007) *The World Food Situation: New Driving Forces and Required Actions.* Washington, DC, USA: International Food Policy Research Institute.

Walter, J. and A. Simms (2002). *The End of Development? Global Warming, Disasters and the Great Reversal of Human Progress.* London: New Economics Foundation.

Warren R., et al. (2006). Understanding the Regional Impacts of Climate Change. Research report prepared for the Stern Review on the Economics of Climate Change, Tyndall Centre for Climate Change Research, Working Paper 90.

Washington R et al (2006) "African Climate Change: Taking the Shorter Route",. *Bulletin of the American Meteorological Society* October (2006) 2006: 1355–1366.

Watts, T.H and S. Watts (2008) "Legal Framework for the Practice of Participatory Natural Resources Management in South Africa", *Forest Policy and Economics* 10: 435-443.

Watts, W.S., 2002. The effects of forestry policy on the sustainability of forest resources in southern Africa. Unpublished PhD dissertation, University of Stellenbosch, Stellenbosch. 298 pp.

WCED (Ed.). 1987. *Our Common Future.* Oxford University Press: Oxford.

WCRP (2007). World Climate Research Programme Workshop on Seasonal Prediction; Position Paper.

Wells M. et al (1992) *People and Parks: Linking Protected Area Management with Local Communities. World Bank,* Washington, DC, 99.

Wescott G. (2002) "Partnerships for Capacity Building: Community, Governments and Universities Working Together", *Ocean and Coastal Management* Vol. 45: 549–571.

Western, D. and Wright, R. M., eds (1994) *Natural Connections: Perspectives in Community-Based Conservation,* Island Press, Washington, DC.

Wheeler, T.R et al (2000). "Temperature Variability and the Annual Yield of Crops" *Agriculture, Ecosystems and Environment;* Vol. 82: 159–167.

White, R. 1998. "Land Issues and Land Reform in Botswana", in *ZERO-REO* Ch. 1.

Wilby, R.L et al (2009) "Review: A Review of Climate Risk Information for Adaptation and Development Planning", *International Journal of Climatology;* Vol. 29: 1193-1215.

Wild, R. & J. Mutebi. 1996. Conservation through community use of plant resources Establishing collaborative management at Bwindi Impenetrable and Mgahinga Gorilla National Parks, Uganda. Working Paper No. 5 of People and Plants Programme, UNESCO, Paris.

Wily, L. A. & Dewees, P.A. (2001). From users to custodians – changing relations between people and the state in forest management in Tanzania, Policy Research Working Paper WPS 2569, Environment and Social Development Unit, The World Bank. 31 pp.

Wily, L.A (2002) Participatory Forest Management in Africa: An Overview of Progress and Issues; DFID Rural Livelihoods Programme, Accra.

World Bank (2002). World Bank *Development Indicators.* On CD Rom. Washington, DC. World Bank.

Wunder, S. (2008b) 'Payments for Environmental Services and the Poor: Concepts and Preliminary Evidence' in *Environment and Development Economics* Vol. 13: 279–297, Cambridge: Cambridge University Press.

Yadama, G.N., 1997. "Tales from the field: observations on the impact of nongovernmental organizations", *International Social Work* 40 (2), 145– 151.

Young, I. M. (1990) "The ideal of community and the politics of difference, in L. Nicholson, eds, *Feminism/Postmodernism*", London: Routledge.

Ziervogel, G. and A. Taylor (2008). "Feeling Stressed: Integrating Climate Adaptation with Other Priorities in South Africa", *Environment Journal.* Vol. 50, No.2: 32-41.

www.ingramcontent.com/pod-product-compliance
Lightning Source LLC
Chambersburg PA
CBHW022315280326
41932CB00010B/1114